Big Rural

Studies in Urban–Rural Dynamics

***Series Editors*:** Gregory M. Fulkerson and Alexander R. Thomas, SUNY Oneonta

This series focuses attention on understanding theoretically and historically the development and maintenance of Urban-Rural Systems through a spatial, demographic, and ecological perspective. It seeks a blending or reintegration of the urban, rural, and environmental research literatures under a comprehensive theoretical paradigm. As such, we further specify Urban-Rural Dynamics as an analysis of human population distribution on social variables, including politics, economics, and culture.

Recent Titles in Series

Big Rural: *Rural Industrial Places, Democracy, and What Next*, by Crystal Cook Marshall

Rural Education History: *State Policy Meets Local Implementation*, by Casey Thomas Jakubowski

Country Teachers in City Schools: *The Challenge of Negotiating Identity and Place*, by Chea Parton

Community Change and Development: *An Urban–Rural Dynamics Approach*, by Gregory M. Fulkerson

City and Country: *The Historical Evolution of Urban-Rural Systems*, by Alexander R. Thomas and Gregory M. Fulkerson

The Rural Primitive in American Popular Culture: *All Too Familiar*, by Karen E. Hayden

Urban Dependency: *The Inescapable Reality of the Energy Economy*, by Gregory M. Fulkerson and Alexander R. Thomas

Urbanormativity: *Reality, Representation, and Everyday Life*, by Alexander R. Thomas and Gregory M. Fulkerson

Rural Voices: *Language, Identity and Social Change across Place*, edited by Elizabeth Seale and Christine Mallinson

Reinventing Rural: *New Realities in an Urbanizing World*, edited by Alexander R. Thomas and Gregory M. Fulkerson

Big Rural

Rural Industrial Places, Democracy, and What Next

Crystal Cook Marshall

Foreword by
Alexander R. Thomas and
Gregory M. Fulkerson

LEXINGTON BOOKS
Lanham • Boulder • New York • London

Published by Lexington Books
An imprint of The Rowman & Littlefield Publishing Group, Inc.
4501 Forbes Boulevard, Suite 200, Lanham, Maryland 20706
www.rowman.com

86-90 Paul Street, London EC2A 4NE

British Library Cataloguing in Publication Information Available

Library of Congress Cataloging-in-Publication Data

Names: Cook Marshall, Crystal, 1971- author.
Title: Big rural : rural industrial places, democracy, and what next / Crystal Cook Marshall ; foreword by Alexander R. Thomas and Gregory M. Fulkerson.
Description: Lanham : Lexington Books, 2024. | Series: Studies in urban-rural dynamics | Includes bibliographical references and index.
Identifiers: LCCN 2023041952 (print) | LCCN 2023041953 (ebook) | ISBN 9781666930740 (cloth ; alk. paper) | ISBN 9781666930757 (epub)
Subjects: LCSH: Rural industries—United States. | Land use, Rural—Government policy—United States. | Rural development—United States.
Classification: LCC HC106.84 .C664 2024 (print) | LCC HC106.84 (ebook) | DDC 338.0973—dc23/eng/20230913
LC record available at https://lccn.loc.gov/2023041952
LC ebook record available at https://lccn.loc.gov/2023041953

♾™ The paper used in this publication meets the minimum requirements of American National Standard for Information Sciences—Permanence of Paper for Printed Library Materials, ANSI/NISO Z39.48-1992.

Contents

Foreword

Alexander R. Thomas and Gregory M. Fulkerson

EDITORS, STUDIES IN URBAN-RURAL DYNAMICS BOOK SERIES

Within the volumes of the Studies in Urban-Rural Dynamics book series, we observe a running theme of rural places, resources, and people being exploited—at a distance—for urban benefit. The urban beneficiaries of rural exploitation are often unaware of their own privilege just as many rural people are unaware of their own disadvantaged location within urban society. This includes the urban consumption of rural water resources, as noted in *Groundwater Citizenship* (Ternes, 2022); the urban consumption of rural food as we considered in *Urban Dependency* (Fulkerson & Thomas, 2021); or, in the present case, the urban consumption of rural energy supplies such as coal. The products of rural work are often encountered in urban spaces as neatly packaged materials on a shelf, hiding and disguising the process that led to their creation. The difficult and dangerous nature of rural work and the environmental destruction left behind by rural production remain unseen by urban populations, who may blame the very rural people whom they exploit for such problems.

For those working in rural occupations—oddly a minority in most U.S. rural areas—the connection to urban is often highly visible and perceived as a source of injustice. Rural workers, such as coalminers, are often quite adept at understanding the meaning of urbanormativity—the tendency to treat urban life as normal and view rural life as abnormal or deviant—even if this is probably not the word that they would use to describe it. In our book, *Urbanormativity* (Fulkerson & Thomas, 2019), we asserted that the vast physical distance between urban and rural populations tends to create greater social and epistemic distance thereby evoking a moral indifference among urban populations for the plight of rural people. In a global economy, physical, social, and epistemic distances are all maximized, and rural workers experience the worst forms of alienation.

In *Big Rural*, Crystal Cook Marshall examines how rural communities are transformed by a large and impersonal scientific system into uninhabitable technical industrial spaces. In one of the most unique and valuable elements of this book, Cook Marshall examines the cultural meaning of industry, noting how it is often imagined in the context of factories, northern mill towns, and urban life. How industry gained an urban connotation is remarkable given the historical origins of many industries in rural areas. Cook Marshall succeeds in breaking down these concepts and reveals their hidden biases and consequences before applying the terms in her analysis.

Cook Marshall also deconstructs cultural notions of rural which tend to be informed by images of White male rural workers that ignore the realities of racial and ethnic minorities who have long played an important role in rural production, including African Americans, Latin Americans, and Native Americans. This is another pattern we routinely observe across the book series and within our own scholarship. Hayden's (2021) book, *The Rural Primitive in Popular Culture*, explains where many of these constructs originate. Misinformed and biased ideas are significant because they have political consequences. When policymakers discuss ways to improve the lives of rural populations, they are limited to thinking of ways to support their White male constituents. Cook Marshall likewise deconstructs the meaning of Appalachia, noting that some of the myths perpetuated through popular culture about rural are intertwined with stereotypes about Appalachia. Hayden highlights this connection, identifying tropes from rural horror films about inbred cannibalistic rural monsters.

As a scholar investigating a rural setting, Cook Marshall looks beyond those rural places in the "commutershed" of urban populations, to places where single economic sectors have prevailed, and where they have often abandoned those whom they temporarily employed. The term "Big" applies to these single economic sectors, including Big Ag., Big Banks, and Big Science. She coined the term Big Rural to refer to the globally connected rural spaces that are drawn upon to fuel modern urban life. Unique to her treatment is consideration of the role of science and technology in constructing Big Rural spaces. Cook Marshall points out that most scholars of science and technology have not given much consideration to issues of rurality, just as most rural scholars have not given a lot of credence to approaches that emphasize science and technology. Her scholarship therefore fills a void found in between two subfields.

Cook Marshall's empirical approach is grounded in field research and interviews. Her primary study site is the Pocahontas Coalfield, where she worked with miners and energy engineers. Her monograph gives voice and agency to rural people, who are often misunderstood or dismissed as being incapable of voice and agency. Even well-intended scholars may be guilty of neglecting to seek out the words of the object of our study: individuals living in rural communities who must deal with a range of social problems that are often absent from or made invisible to urban society. For example, Cook Marshall, offers longer sections of perspective from three rural coalmining community advocates. In those, respondents express a regionalization of urban-rural systems, which is consistent with ideas expressed in the recent volume, *Community in Urban–Rural Systems*, by Fulkerson (2022).

In addition to interviews, Cook Marshall examines documents on safety, science, economics, the environment, and epidemiology. She attended professional conferences from different perspectives, including those that espouse a pro-fossil fuel stance. These efforts enabled a more nuanced understanding of and compassion for regional actors connected to the study area. She also discovered a rift between different types of actors and stakeholders that prevented communication. With different constituents operating in different silos, she reveals an easily remedied barrier to communication.

The takeaway from Big Rural, as noted in the Introduction, is that the fossil fuel and other natural resource extraction industries exert undue influence in constructing regional culture and space, and their primary tool for accomplishing this is science and technology. Most rural scholars studying Big Rural issues (namely, Big Ag) have done so from a political economic perspective heavily informed by a Marxian framework. While Cook Marshall acknowledges the important role of capital in funding these Big Rural industries, she rightly points out that capitalists operate at a great distance, exerting almost no direct influence on how businesses operate on the ground. The ones carrying out the transformation are part of a more proximate web of actors who create changes by using science and technology to flatten rural spaces into large industrial laboratories. These transformations result in severe social, cultural, economic, and environmental consequences, which can be highly damaging and detrimental.

As the editors of a series that often reports unfortunate patterns, we appreciate when there is some glimmer of hope for improvement. Cook Marshall provides guidance on how to improve the situation by highlighting a path through or around technocratic transformation. No band-aid, it will require us to rethink and reorganize the current structures and systems. This will require support from federal and state actors, who are themselves often intimidated or controlled by Big Rural interests. To address the problems of Big Science, Cook Marshall advocates a move toward citizen science that can alter the scientific institutional structure dominated by large university engineering programs as well as adjacent programs in geology, earth and mineral science, and even fishery, wildlife, and forestry programs. Promoting citizen science returns a measure of democracy to an otherwise technocratic top-down governance model.

We welcome this addition to the book series and trust that the reader will find a great deal of value in the pages that follow!

Preface

Mode of Inquiry

Starting in 1967, Ukrainian-Brazilian writer and journalist Clarice Lispector contributed a weekly column to a popular newspaper in Brazil. In one among those collected, Lispector interviews Chilean poet-diplomat Pablo Neruda and poses to him the question of what the most important thing in the world is. Neruda responds, "To try and make the world a worthy place for everyone, not just the privileged few."[1] If there is an overarching aim for *Big Rural* it is to reveal less investigated research and historical, theoretical, analytical, and practical paths toward assisting with making rural spaces places worthy for everyone. In this volume, I likewise take seriously the advice of scholar Louis Menand that the main goal of an academic in a democratic country ought to be to investigate lines of inquiry the greater public may not want to have asked and to seek answers to questions the public may not be able to ask. Menand encourages scholars to "accommodate" "the voices it [the public] fails or refuses to accommodate." For Menand, the goal of academics ought to be to look beyond themselves and reflect on what sort of scholarly work and teaching the world needs, then to go do that, yet, at the same time, to also "ignore the world's insistence that they reproduce its self-image."[2] In this regard, with *Big Rural* I seek to stand as a critically and regionally engaged scholar who also participates in a place of inquiry.

Though I demonstrate an ability to step outside my personal experience of place per Menand's call, I do not assume a nonidentity. At our juncture in history, substantial work in many fields occurs in which scholar-activists or scholar-participant observers include themselves as part of the scholarship. For example, in Science and Technology Studies of Society (STS), in the classic work *The Structure of Scientific Revolutions*, Thomas Kuhn traces his maturation in thinking leading to his development of the concept of paradigm shift. In his book *Critical Regionalism: Connecting Politics and*

Culture in Appalachia, regional and Appalachian Scholar Douglas Reichert Powell analyzes his own childhood and class in Knoxville, TN, using these as a springboard on the issues of the academic and place; likewise, in the work of Talmage Stanley on the Pocahontas Coalfield, he recounts his family's history as central in his examination of the cultural and economic fabric of Appalachia in the early twentieth century.[3]

I begin with a personal narrative, largely included in this *Preface*, and then move to a macro analysis moving toward filling in the description of Big Rural with spaces that constitute a type, a kind. This move pulls the arc of this volume away from the sometimes-intimate gaze of place in Regional, Rural, and Appalachian Studies toward analysis of how rural industrial spaces function in total (Big Rural) in society, and, what role science and technology (read engineering) plays there, and what it ought to play now, given the historic outcomes of the rural industrial space in terms of economy, social problems, and environment. Rooting this first step in the personal situates the urgency of the scholarly treatment at hand and charts a process of moving from visceral impulse to sustained academic insight.

In terms of this personal journey, I also weave into this expository text three interludes of personal narrative/voice from other people from this region—coproducers of economic sector change in the Southern West Virginia/Southwest Virginia coalfields. Not only do their voices situate the issues in the rural now, but these participants assert their own knowledge. In the inclusion of these interludes, I also acknowledge that localized knowledge, referred to by scholar James C. Scott as metis, is often extracted by academics or captured by makers of media for their own use. Both in form and practice my work in the rural and on Big Rural seeks to reconstitute local metis and to assert its value.

These interludes serve as points and spaces for reflection that this narrative/exposition, while written by me, is not mine only, and, they give the reader an opportunity to draw throughlines from historical or theoretical considerations to the concrete work on the ground of seeking and working for tangible change, and, the stakes leading to and propelling someone to work for that change.

OVERVIEW

In the United States, we assume the rural space as backward and unsophisticated and as a space devoid of technological intervention. *Big Rural* demonstrates how science and technology demarcate and prescribe the "rural" space in the United States just as much as they create and contribute to the performance of metropolitan/sophisticated spaces. Popular culture sells the rural space as wild, natural, and untouched; the reality remains that the United States has been thoroughly categorized in geography and topography for its utility for science and resource exploitation. Much of our "wild" now consists especially of plants and insects from around the world: very few old growth forests remain. We cannot return to pre-Columbian North America. Both urban and rural exist on a continuum of what science and technology prevail in each space.

I present a case study, accomplished through interviews, archival sources, participant-observation, and literature review, of how a rural space came to be in use as a coalfield rural industrial space. From these resources and from theory, I extrapolate broader lessons about the U.S. rural industrial space toward a conceptualization of the facets of a Big Rural. I argue that the science and technology responsible for enacting the rural industrial space flattens the environment, dominates education, diminishes knowledge accumulation, guts on-the-ground democratic practice, leads to automation displacing people cum technology, and contributes to the rural industrial being black boxed in the larger culture, its wares and raw product part of the magic of modern life's offerings simply appearing (food in stores, energy in the socket, water in the faucet, gas at the pump—the invisible ubiquitous fruits of Big Rural). I proceed to substantiate the drawbacks of this rural industrial model, or, strategy void in U.S. policy.

I challenge the scientific and technological imaginary for the rural industrial: these spaces call for fresh science, new research, and new scholarly and theoretical engagement. They call for a reconfiguration of science away from corporate and crony sponsorship toward a socially responsive and responsible science. I enumerate what a U.S. rural policy ought to address in this space and review democratic and nondominant economic practices arising in the post-single-sector rural industrial space.

I also reflect on what an STS examination of the rural industrial space gains us in terms of understanding the space's role in society and conceiving of its relationship to research, and technology and democracy.

PERSONAL REFLECTION

Don't send me another treatise labeling White Appalachian Mountain people as some special breed—neither in their degenerate behavior nor in their nostalgic family fealty. Don't send me strumming rehashes of their connection to a pre-coal purity, a pristine time of childlike ways, when no one's White Appalachian mountain or lowland ancestors lorded over house or field enslaved peoople, or, in equally brutal times prior, cut the throat of an indigenous person, or at least accepted the land grant of those that sanctioned that doing, taking, and staking of a sliced and sectioned off land, now fondly remembered as the "old homeplace."

I am not here to defend anyone, not coal miner, not landowner, not coal wrangler, not land holding company, not mountaineer, not investor—and not scientist. I am not writing this to show that I matter, or my family's history matters, as much or more than anyone else's, because my family was good people, after all. I am not writing this to remove blood from my hands or absolve anyone or myself of anything.

I am not writing to redeem. Reflect, yes. Rethink, I hope so. Rescue? I am not writing to rip from anyone their agency.

The people remaining where I am writing about, if adult, are people who have been handed and handled plenty of life experience, and can make their own decisions—whether those were to vote for Mondale (Democrat) in 1984 or for Trump (Republican) in 2016. They, too, are as complex, honorable, and flawed as anyone anywhere else.

But not only are people at stake in Appalachia—there is the land, the water, the air, too. In full disclosure, I want you to know this:

I love the rusty, creaky, grimy, fallen down, apocalyptic Appalachia—I grew up in a post-employment coal field in Appalachia until I was ten years old, then we moved from there to "Cancer Valley" or the Kanawha Valley, full of miles and miles of chemical factories. Yet, in my growth as a scholar, I have come to question my own nostalgia for place, and my own acceptance of created boundaries of space. Prior to my doctoral work, I was certain what Appalachia was, what counted as Appalachian, and that being Appalachian was demonstrable, categorizable from other demarcations of space. Now, while I may lament for a specific mountain, I cannot honestly wring my hands over my local site and locale without seeing its connection to spaces across the country, and across the globe, because what happens in my space of close examination—the Pocahontas Coalfield of Appalachia—happens elsewhere. To only say WE from there count as effluence misses the flow and stream we are in with so many people and places. So, while I make my own yard my subject, it reflects the off-shored, out-of-sight, out-of-mind rolled-over places being sliced into the earth's sides worldwide to mostly feed societies' centers (aka Big Rural).

Discussing Appalachia, I am reminded of a friend's anecdote of Russian (someone else told me it was generally Slavic) machismo—-a Russian man sits at a table drinking vodka, smoking a cigarette; he moans, "Russians—we are shit. We have always been shit. We will always be shit," while behind him a woman busies herself making dinner, looking after the children, getting things ready for everyone and herself for the next day of activity—school, work. Writing and reflecting on Appalachia strikes me sometimes as that kind of indulgence—a marginal luxury of practicing fatalism, while those adjacent to us there, or left there by happenstance or choice, focus on the now and the next day and keeping what show there is left, running.

While I sometimes get down in the weeds of this Appalachia, where I have been and where I am, and while I can appreciate Hemingway's advice to not write about Man, but a man, and while I can also appreciate that the work of legitimizing particular peripheral perspectives, places, lives is endless work, that is not where I arrive here. Having also worked in the particulars of literary fiction—reworking and representing mostly Appalachia in that genre, as well as in creative nonfiction, I am a little worn out by the up-close, intimate gaze into Appalachia or, the first-person authority, the melding and meshing look through Appalachian eyes (though, I engage with some of that *in dialogue*, in interview, with some local agents for change as interludes, but not asides, to the theoretical throughline of this text). Some of the story, essential story, seems to be missing in the box now most often drawn of that place—a box that is literary, poetic, cultural, anthropological, sociological, and political economic. Hence, the launchpad for this current examination is STS: sociology of science and technology, history, philosophy, policy, and science and technology culture. My hope is that this path brings me, as a scholar activist, and community practitioner, beyond practices of, and indulgence in, fatalism—past smug assumptions of lack of agency on the part of regional participants. Is the woman in that Russian anecdote above smart, more adept for taking care of the work before them both, or is she complacent, stupid, acquiescent, cajoled, cowed? Does everyone else depend on her more than we can imagine? In any case, she gets up and keeps at it—and even in the best of times, existentially, something can still be said for that.

In 2002, I took a brief road trip back to the county of my birth (in the flyover, we talk counties—we assume you probably haven't been to the specifics) to the top of the Pocahontas Coalfield in Southern West Virginia. By then, the area had slid even further down economically in terms of individual employment. In addition to following the mostly then empty main streets of Princeton and Bluefield, West Virginia, with their abandoned storefronts, deteriorated downtowns with once stately or even artistic buildings, falling in on themselves, I also drove down to Bramwell, a Victorian hamlet, tucked back from Route 52, and that route's mix of blown-out trailers, long gone businesses, and derelict houses. Bramwell, by comparison, registers almost fairytale-like. As the Pocahontas Coalfield was being wrestled into an industrial occurrence from the mid-1880s to the 1920s, Bramwell was one of dozens of locations created whole cloth by the incoming industrial wranglers. Usually, Bramwell is discussed academically and locally as having been inhabited by coal "barons" and having once been the richest per capita place on earth: once these barons had "opened" up access to the coal seams of the region, they made their fortunes and built this stately town. Instead of baron, I employ the term "wrangler" here, as the word baron implies landed gentry not engaged in dirty work. By contrast, certainly a number of the men, and at the level of industry rule they were all White men, who engaged in the activity to "open" up this particular geological set of coal seams to export, were far more cowboy than count, and, more speculator and prospector than a spectator. Bramwell is the town where they invented themselves into a local aristocracy, despite having wrangled, wrestled, shafted, and wrested industrial production into being in a short time in a place often described then more as a remaining Eastern version of the American West. Prior to and in the early days of timber and coal, civil rules and laws took a backseat to those of might, and homesteaders eked out livings in a rough territory and propped themselves up in other ways through livestock sales and hunting and gathering, bartering their cultivated or gathered plants, herbs, or animals into trade. Thus, Bramwell, a prim town, replete with mansions, high society gatherings, and cosmopolitan tastes, was an odd and recent development into this otherwise frontier, and, also a quick-to-rise and quickly raised symbol of assertion of dominance by the wranglers over the territories they had recently conquered in the name and the game of coal.

How had these wranglers accomplished this, where were they from, and why did they do it? Largely men who had attempted building wealth in other industrial areas, many sought their fortunes in the Pocahontas Coalfield, bringing with them experience or family legacy in the Pennsylvania Coalfield (opened some hundred years prior), and from the Welsh coalfields of England prior to that.[4]

Back in 2002—I returned to the county of my birth for a brief visit and, although I lived twenty minutes from Bramwell from the time I was born until I was ten years old, I had not ever visited. My childhood had been peppered with countless rides past Bramwell as we traveled Route 52, hairpin turning into the coalfields to visit family friends in McDowell County's county seat of Welch. Often, too, on those visits, my father would veer off a road and up a hollow to point out some site of family legacy: here was the house in Leckie where he was born, and look, a light is on. Someone is even still living in it! Those were some of the coke ovens your grandpappy assisted Italian masons with as a water boy after he'd completed the sixth grade—sixth grade was as many grades as there were back then. This is the section where Cuz (our cousin Jim) grew up in Keystone. That road was where I worked as an insurance investigator for Equifax (my father worked there during the 1960s and early 1970s—it was one of two non-coal jobs listed in the Bluefield, West Virginia newspaper when he got out of the Army in 1961, the other job listed was selling Fuller brushes) and I encountered XYZ (usually some sordid tale). That road there took you to Matewan, where your grandpappy had also fought for the union. This company town is where your Uncle Junior met Madelyn (my aunt) and her Hungarian parents lived. But that town there is where Junior fought for the union. And this place here is where Junior had his own little pick operation, a mine he worked for several years with a few other guys doing by hand what big machines were doing by then. That mine was no longer worth scraping at for a big company, so they sold it off to people who could still maybe make a living off it with carts and mules. Working that mule one summer cured me of ever being in the mines. Oh, and this town is where the Black folks lived. And this town here. And this one there. These are where the White folks lived. This is where your grandmother got off the train to come visit when she took the rail from Elliston, Virginia, near Christiansburg, to Pageton. Imagine—you could ride a passenger train back then to almost anywhere big enough to be given a name!

Look, see that outhouse along that bank there heading out toward that creek, what comes out of that will eventually end up in the Ohio River. Same as if you use the bathroom in that fancy house—most everything here is straight-piped. Now, this is where the Longos lived in Welch, and that Catholic Church is where they go to church still. Joe Longo is who taught your grandpappy to make real pizza and real spaghetti sauce. Those scrubby pines along the hillside is where and what they have planted back after strip mining. And right here now, comes Mr. Peabody's Coal train, a Norfolk-Southern one mile long, fully loaded, winding its way around the bottomland, along the river. You know?

No one down here owns that. All the bigwigs in this—they are all up North. Your grandpappy would often recite this:

> *Here is to Boston, to the baked bean and the cod,*
> *Where the Cabots speak only to the Lodges*
> *And the Lodges speak only to God.*[5]

Later I learned that a cousin of mine, an accountant, owned one of the former coal wrangler homes in Bramwell for a time in the 1980s, yet we never did venture over that way. Looking back, I can either assume that my father took for granted that since my family had not worked in Bramwell, he had little to say about it, or, as my father had little good to say about the wealthy and/or managerial class of people of the region, he might not have had any interest in gawking at their displays. Why stare up at those on purpose who had put their foot down on your family's necks?

Yet, I turned off Route 52 in 2002 and entered this fairy tale town. As I said, the main route leading up to the turn off had dilapidated houses, empty gas stations, run down trailers, and blown out buildings, most recently affected in 2002 by an influx of meth labs and other illegal commerce. Crossing over into Bramwell, with its neat small streets, its dozen or so visible Victorians, and its smattering of still standing brick facade main drag buildings, I kept thinking I had crossed into New England. Or maybe one of those Victorian neighborhoods I had seen in California. It also struck me then that this little hamlet would be a lovely college. Wouldn't that be something, I thought, a lovely college right here, I wondered, or maybe a retreat. Or, with its views and serenity, an art college? West Virginia had no standalone art institute. Wouldn't that be crazy? An art college right at the mouth of the Pocahontas Coalfield. My father had avoided mining coal by becoming a self-taught graphic artist and teaching himself Benjamin Franklin's profession: printing. Imagine, I thought, all the things that an art college right here would bring!

I shelved this concept as something that would be neat but did not actually consider it to any real degree until 2006. That year, after ten months in Armenia on a creative writing Fulbright, I returned to my then city of residence, Los Angeles, finished the makings of a divorce, and sat with myself thinking, what else could I be doing? This concept for an art school in the coalfields came back to me, and I reached out to several sets of folks for discussion. One, Phil Hanes, who sparked the founding of the University of North Carolina School of the Arts, willingly wrote back and forth with me for a while per email, advising me to reach out to Sharon Rockefeller—the wife of Senator Jay Rockefeller. That contact and one made for me by Jeff James of a group called Create WV to the Claude Worthington Benedum Foundation maybe a year or so later for advisement resulted in dead ends. At that time I put together an outline of what this kind of school might focus on: rural arts not only for the Appalachian region but a place for rural focused art and creativity from rural spaces around the world. In the meantime, I started a job as a project manager and then executive director of a nonprofit project of the actress Geena Davis to improve gender portrayal and representation in content aimed at children. I shelved this concept for a rural art school and moved on with this work in Los Angeles.

By 2009 I found myself again sitting in West Virginia, thinking. about what I ought to do with the second half of my life. This concept of a School of the Arts in the coalfields returned to me. By then, much ado had been made in the United States both in the popular press and intellectually about the New Creative Economy and its purported benefits for places in economic decline. Most of this work had been focused on large cities, but I wondered if this solution of arts influx might function as promised also in a more out-of-the-way place. With respect to all of this, at that time I had far more questions than answers. Moreover, I had no idea how to start a school, much less whether a school would make any sense at all to do where I thought it seemed a good fit. What did one study to find the answers to this?

Casting a wide net, I applied to Ph.D. programs as disparate as Digital Media (as at that time, like many people, I conflated the term technology with computers, and, understood technology to be good, its arc positive, and it as the driver of economic progress), Education (but those programs largely focus on secondary schools), and something I came across called "Technology, Innovation, and Policy (TPI)"—as those three all seemed reasonable fields to examine if starting a place of education in the twenty-first century. A friend already well-established in academia cautioned me—with respect to which PhD to pursue, to "go with the one that pays for you." August found me and my newish second husband in a graduate dormitory in Stony Brook, Long Island as I delved into TPI classes in a school of engineering. There, while in a class on the philosophy of technology, I encountered the field of STS. By the next year, in 2011, and with a National Science Foundation Graduate Fellowship in tow, I relocated to the STS department at Virginia Tech in Blacksburg, VA. By then, I had been disabused of the concept that technology was largely only digital and that this, or any other, technology necessarily prompted positive economic or other positive change. After additional academic and research detours into energy, science, technology, and rural policy, and Appalachian and Rural Studies, along with much personal work alongside my husband on a farm we had in North Carolina (a couple of hours from Blacksburg, VA), after engaging in in-depth interviews with a range of grasstops (people heading up grassroots and other civic and entrepreneurial engagement) in the region, and after attending many regional meetings and conferences as a participant-observer, I found myself in 2016 coming back full circle to some kind of school, some kind of institution, as a necessary kind of invention, an undulation, to ripple the economic tide in the Pocahontas Coalfield.

In short, a whole economic system had been intervened in this region with the advance of industrial-scale coal production in the mid-1880s. The employment run on this had been reasonable, given commodity fluctuations, until the introduction in the 1950s of a machine called the continuous miner, which allowed coal companies to cut their workforces in half by 1960. For much of this history, the United Mine Workers of America (UMWA) acted as the organizing force, the social capital institution providing social coherence for the miners. In the 1980s, with the larger push nationally to downgrade the capacity of unions as a countervailing force to corporate demands upon workers, the UMWA also lost much of its remaining potency.[6] Thus, along with the continued free fall of jobs in the coalfields due to technological advances, the union also weakened as an institution around which to rally.

In September 2016 I attended a meeting of the Central Appalachian Network (CAN), a regional network of nonprofits that works together on economic issues. At that meeting, during a session on the new food and sustainable agriculture economy in the region, one table I worked with made a map of the anchor or other institutions working on that throughout central Appalachia. Glaringly apparent was the lack of an organizing institution of this sort in the Pocahontas Coalfield and its immediate adjacent counties. Either no space had been created because that emerging economy had no possibilities in the Pocahontas Coalfield (which I knew to be untrue), or we had been looked over, passed over, left out. This meeting confirmed for me that brick-and-mortar institutions can serve as the lynchpin for the public good: in this case, supporting land-based (agricultural, botanical, forest-based) production capacity.

Long-standing examinations in economics and economic geography review the interplay of geography, strength of institutions, economic development in extractive regions, and variables of trade. In the U.S. context, the strength of the civil society sector in a region may also indicate the strength of local democracy.[7] Civil society is an integral part of democratic society, and where essential civil society institutions are missing, then there could also be a weakness of local governmental institutions in terms of community trust.[8] Rodrik, Subramanian, and Trebbi offer strong evidence for the primacy of institutions in determining economic well-being.[9] However, their work does not prescribe how to change governments or civil society to become more robust.

Though previous to this CAN meeting I had flirted with the academic literature on the importance of actual brick-and-mortar institutions, at the meeting itself I could clearly see their functions as nearly technological. The shape and robustness of civil society as a contributor to local democratic practice could also dictate what was possible on the ground. The lack of the latest technology could leave a region behind economically, but not necessarily if other factors like civil society are robust; likewise, a technological fix will not necessarily improve the economy if civil society and democracy are not also robust.[10] For example, there has been much activity and push for the need for broadband in the coalfields or in rural areas more generally. However, broadband (or the next major technology after it) should not be an end stop—without other robust democratic organizing principles or institutions and kinds of technical or technological intervention, broadband or anything else even of more consequence might not amount to much.

While working for the energy sector in 2015–2016 I had a chance to review what constructed a robust economic sector or at least a resilient one in the region at hand. Though far from its peak heyday, in recent years, even with its decline in domestic use, coal has continued production and use, and, export from the Appalachian region.[11] Dozens of macro, policy, local, and social factors construct its production feasibility, and, from examining a handful of these up close, I assumed that also dozens of these factors would construct the sustainability prospects of any other economic activity of scale or scalable in, or adjacent to, the Pocahontas Coalfield. To this, I add an assumption that like the coalfields, other economic sectors function similarly with much broader implications, impact, and participation beyond the profession most associated with that sector. Professions like "coal miner" or "farmer"—professions nearly or also having an archetype are connected to or somehow absconding dozens of more professions and people also deriving their livelihood from these sectors through research, sales, logistics, equipment, fabrication, construction, and so on.

In this work, I engage most demonstrably with Langdon Winner in his examination of the totalitarian workplace and what this means in a single economic sector space. I examine the rural industrial space as a large technical system and how science and technology have enabled a flattening of the environment along with community disintegration and destruction of people's livelihood, and, contributed to knowledge deterioration.[12] I reflect on how policy only matters if you have institutions capable of enacting them. I return again throughout the text to the problems in the twentieth-century rural industrial space when people as technologies are finally replaced by machines, and, specifically, what automation of once human-done work has wrought in the Pocahontas Coalfield region. From here, I posit cooperative economic and community-oriented technology policies as ethical responses and reflect on enacting a brick-and-mortar institution as a means for economic shift.

POCAHONTAS COALFIELD AS SUBJECT OF THEORY AND ANALYSIS

Much discussion of Appalachia, and the Pocahontas Coalfield in particular, begins well after the introduction of active industrial coal mining. Both popular media and academic examinations of the region dwell on this and the region's persistent poverty.[13] In much the way that writer Binyavanga Wainaina in his essay "How to Write About Africa" compiles the stereotypes that pepper popular and other depictions of that continent, before working through analyses of the Pocahontas Coalfield, I wanted to confront two related issues: Appalachian degradation and Appalachian exaltation.[14] These two features of the discourse, both by insiders and outsiders, can fall into truisms and points of contention by both camps. On the one hand, what I call the "Daddy's on the porch playin' fiddle while Momma's in the kitchen singin' and bakin' biscuits and I sit in this rockin' chair, patchin' a quilt" syndrome requires the author or commentator to exalt the "true" Appalachian "virtues" of close families, traditions in foodways, handicrafts, and music—freezing "authentic" Appalachians in time, in race, in class, and in pastoral close-knit, loving families and communities. Left out from this "true" Appalachia, both by insiders and outsiders, obviously, is anyone whose experience does not mold to these conventions, thereby erasing a significant swath of people and places (e.g., urban Appalachians, coalfield Appalachians, industrial Appalachian, etc.). Layered onto these "true" depictions include tales of virtuous working-class men, whether small farmers or miners, their loyal women, and the willingness of offspring to look out for each other. The contrast to this depiction includes what almost invariably comes out of a new acquaintance's mouth if this new acquaintance is (a) from the East Coast of the United States and (b) not from Appalachia—that is, a reference to incest, though, I would say that since the mass media reporting and documentary filmmaking about the Opioid epidemic circa December 2014 (on the backs of reporting by Ken Ward of the Coal Tattoo), addiction may be displacing incest as the popular media or outsider association.[15] These extreme narrative depictions by outsiders become more layer in terms of exploration of why West Virginia has been allowed to become so environmentally degraded.[16]

Both the West Virginia narratives of exaltation and of degradation lead insiders and outsiders, albeit on quite different grounds, to contend that West Virginia, or Appalachia as proxy or vice versa, is exceptional—unlike anywhere else in people, circumstances, and place. I contend that this exceptionalism contributes to the potential for West Virginia's, and Appalachia's, further degradation; this exceptionalism bars West Virginians or other rural peoples from seeking folks in similar circumstances in the United States and worldwide with whom to form solidarity or seek affinity. In my fieldwork and participant observations of grasstops, some mission-driven regional nonprofits arbitrarily erect boundaries in their mission or work at the created "Appalachian" border—assuming Appalachia as a fixed space to the extent I came to question the utility of Appalachia as a concept in its post-coal or post-small farm twenty-first-century state. Much deemed Appalachian, with examination through Rural Studies, can nearly be pointed out to be located also in other rural American places, from foods to self-reliance myths to music traditions.[17]

Moreover, what additionally prodded my suspicions in the case of both coal and poverty and the narratives embracing their intertwinings was a question that many narratives seem to have shadowing their content, lurking in the background: what is the relationship between Appalachian coal production and Appalachian poverty? Likewise, cultural and family myths have built up on the same themes of why people stay though poverty persists, spouted in phrases like, "These people will never leave. They have mined here for generations." Or, in turn, other myths of ancestral right and bravery with respect to the local land's defense: at a conference on "Social Enterprise" in 2016 at Marshall University in Huntington, West Virginia a speaker proclaimed her family's fight against surface mining—they did it because this land was her granddaddy's. Her take: her family's historical ownership was more authentic than what a coal company could lay claim to. From this as well as other contemporary discussion on coal and poverty, I tease out the coal + poverty backgrounded narratives that go something like this:

> *Look at this mess—it's an environmental and social nightmare. Just why on god's green earth would anyone mine coal? What is the matter with these people that they would do this or live here?*
>
> *People are so poor here. Why do they stay? You must be awfully damn dumb, corrupt, hopeless, ignorant, to stay or to have even come here at all.*
>
> *Well, look at them. It has always been this way. These people, this coal; these people, these mountains.*
>
> *My claim, as a White person with ancestors who settled these hills, regardless of how I came to be in them, my claim is stronger, more authentic, more caring, more nurturing, more thoughtful, by virtue of my longer claim and ownership.*

Add to this the folks in Appalachia who construct themselves as outsiders to the principal coal regions, highlighting their own sophistication by contrast, stating about the citizens of the coalfields, "Those people will never change."

NOTES

1. Clarice Lispector and Giovanni Pontiero. *Selected Cronicas* (New Directions Pub. Corp., 1996).
2. Louis Menand, *The Marketplace of Ideas: Reform and Resistance in the American University*. New York: W.W. Norton, 2010.
3. Case in point, my first experience with this stance was in a class at Columbia University in 1993 with the anthropologist teaching it and also writing about his affecting his subjects while doing fieldwork in El Salvador. Douglas Reichert Powell, *Critical Regionalism: Connecting Politics and Culture in an American Landscape*. Chapel Hill: The University of North Carolina Press, 2007; Thomas Kuhn, *Structure of Scientific Revolutions*. Chicago: University of Chicago Press, 1962; Stanley, Talmage. "The Poco Field: Politics, Culture, and Place in Contemporary Appalachia." Emory University; ProQuest Dissertations Publishing, 1996.
4. Fred Brenckman, Official Commonwealth Historian (1884). *History of Carbon County Pennsylvania* (Harrisburg, PA: J. Nungesser 1913).
5. John Collins Bossidy, Toast at the 1910 Holy Cross College Alumni Midwinter Dinner.
6. Shannon Bell, "'There Ain't No Bond in Town Like There Used to Be': The Destruction of Social Capital in the West Virginia Coal Fields." *Sociological Forum* 24, no. 3 (September 2009): 631–57; Rebecca Scott, *Removing Mountains: Extracting Nature and Identity in the Appalachian Coalfield* (Chicago: University of

Minnesota Press, 2010); David Harvey, *A Brief History of Neoliberalism* (New York: Oxford University Press, 2007).

7. Marc Morjé Howard, *The Weakness of Civil Society in Post-Communist Europe* (Cambridge, 2003).

8. Bell, "'There Ain't No Bond in Town Like There Used to Be," 2009.

9. Dani Rodrik, Arvind Subramanian, and Francesco Trebbi Source. "Institutions Rule: The Primacy of Institutions over Geography and Integration in Economic Development," *Journal of Economic Growth* 9, no. 2 (June 2004): 131–65.

10. Ibid.

11. Analysis. "Appalachia Comes Up Small in Era of Giant Coal Mines." Washington Post, May 5, 2017. https://www.washingtonpost.com/graphics/national/coal-jobs-in-appalachia/?utm_term=.3399264b6955.

12. Landon Winner, *The Whale and the Reactor: A Search for Limits in an Age of High Technology* (Chicago: University of Chicago Press, 1986).

13. Emily Satterwhite, *Dear Appalachia: Readers, Identity, and Popular Fiction since 1878* (University Press of Kentucky, 2015); Douglas Reichert Powell, *Critical Regionalism: Connecting Politics and Culture in an American Landscape* (Chapel Hill: The University of North Carolina Press, 2007).

14. Binyavanga Wainaina, "How to Write about Africa." *Granta*, no. 92, Cambridge, 2005; For more extensive discussion, see these academic, literary, and non-fiction references dealing with the topics of activism, historical political or ethnic subjugation, mining, or perceptions of the people of Appalachia. For example: Rodger Cunningham, *Apples on the Flood: The Southern Mountain Experience* (Knoxville: University of Tennessee Press, 1987); Ann Pancake, *Strange as this Weather Has Been: A Novel* (Shoemaker and Hoard: Emeryville, CA, 2007); Huey Perry, "*They'll Cut Off Your Project*"; *a Mingo County Chronicle* (New York: Praeger Publishers, 1972); Barbara Ellen Smith, *Neither Separate Nor Equal: Women, Race, and Class in the South* (Philadelphia: Temple University Press, 1999); Joe William Trotter, *Coal, Class, and Color: Blacks in Southern West Virginia,* 1915–1932 (Urbana: University of Illinois, 1990); Frank X. Walker, *Affrilachia* (Lexington, KY: Old Cove Press, 2000).

15. Murray Energy used this excuse to explain coalfield illness, that is, "poor health due to incest." "Dr. Michael Hendryx, Measuring Mining's Toll on Health." PRI's Environmental News Magazine. Living On Earth. Boston, MA: Public Radio International, August 11, 2012. http://www.loe.org/blog/blogs.html/?seriesID=1&blogID=17.

16. Crystal Cook, "Replicas," southernhum.com, 2006.

17. David L. Brown and Louis E. Swanson. *Challenges for Rural America in the Twenty-First Century. Penn State Press*, 2010; Schafft, Kai A., and Alicia Youngblood Jackson. *Rural Education for the Twenty-First Century: Identity, Place, and Community in a Globalizing World.* University Park, PA: Pennsylvania State University Press, 2010; Campbell, Hugh, Michael Mayerfeld Bell, and Margaret Finney, eds. *Country Boys: Masculinity and Rural Life*, 2006; Duncan, Cynthia. *Worlds Apart: Why Poverty Persists in Rural America*. New Haven: Yale University Press, 1999; Kenneth Johnson, "Demographic Trends in Rural and Small Town America." The Carsey School of Public Policy at the Scholars' Repository, March 15, 2006. http://scholars.unh.edu/carsey/5.

Acknowledgments

I am very grateful to a range of people who have made my work possible. I am extremely grateful to the National Science Foundation for their support for three years. I am grateful to the community of scholars at Virginia Tech for their feedback and attention and the late John Craynon for allowing me insight into scientific and technical production in the region. Many thanks also to these additional folks who have provided input, ideas, and solid approaches for my evolving research, outreach, shaping of policy potentials, and on-the-ground development work.

With particular thanks to:

- The Studies in Urban-Rural Dynamics and Lexington Books
- James Collier and the Science, Technology and Society Department at Virginia Tech
- Harry Hamil, Morning Chorus Farms
- Jason Tartt
- Amelia Bandy
- Shannon Lenahan
- The inimitable Atlas Charles
- The Department of Technology, Policy and Innovation at Stony Brook University
- The folks in and associated with the Eastern Regional Coal Archives at the Craft Memorial Library in Bluefield, WV
- The continued engagement from the folks in the Town of Kimball, WV
- My husband for his encouragement and for doing what had to be done for us as a family so that I could pursue this research, and, for sharing his depth of knowledge of agricultural practice and his unwavering ethics with

respect to that practice. I am eternally grateful for your support and our work together in farming.

- Allynne Perkins for his incredibly generous proofing of citations and sources

And the many other people who have let me interview them, pester them for information or direction, attend their meetings, go to their conferences, or have pointed me and engaged me toward resources or intellectual direction. Thank you!

Introduction

In the early 1990s, I lived and traveled in the Soviet Union, and what I saw there reminded me of the deindustrialized landscape where I grew up in and around the coalfields of West Virginia in central Appalachia in the United States. Similar in kind to the abandoned waterfront of Brooklyn, New York where I moved to find cheap rent a couple of years later, rather than see all of these places as far apart, I saw their links: communities created whole cloth to serve industry and then abandoned once the physical labor of the human inhabitants was no longer needed, off-shored to cheaper human labor, or, automated. Having lived long enough to see that abandoned waterfront in Brooklyn sucked into the real estate booms of the twenty-first century and those warehouses converted into or replaced with luxury apartment buildings, I also worked and traveled during those same years in the former Soviet South Caucasus—urban and rural Armenia and the Republic of Georgia, and, also in central Appalachia, in West Virginia, Southeast Ohio, Eastern Kentucky, and Southwest Virginia. In these places I saw the ravages of single-sector economic myopia of both communism AND capitalism, and, again, I saw the links.

I originally conceived of a research project that would take me to all these places to research and consider these links, but then, like maybe we may so often do, I found more than enough to analyze and reflect on in my own place of origin—the Pocahontas Coalfield of Southern West Virginia and Southwest Virginia. Landing in the field of Science and Technology Studies of Society (STS), the theories offered me there of the socio-technical, and the language of laboratories, coproduction, and large technical systems (LTS), gave me terms I could lend to what I witnessed and researched with respect to the creation of a rural industrial space like a "coalfield," and also lent me concepts I could bend to describe the scientific and technical expertise and experts and their roles in coproducing the knowledge that lead/led to the creation of such a space, and, to its continuation.

A reviewer of this volume commented on what they viewed was a potential excess of jargon (I have since worked with lay readers to put that in check, largely by explaining jargon rather than eliminating it) and part of my initial project for this book was in finding the terms to explain the connections among these rural industrial spaces that I saw, and, which seemed obvious to me, but obscured in terms of academic, field-specific theory or examination. Originally a fiction writer and a writer of historical fiction, I had spent a lot of time in my twenties and thirties putting to the page stories that seemed obvious to me, but invisible in the larger culture: modern family dramas in Appalachia, stories unfolding with deindustrialized landscapes as their backgrounds, stories of the complexities of race, gender, and class, stories of loss of hope and of strong drugs and broken dreams.

Wading through and sometimes drowning in terms, about the time I decided to devote my energy to understanding how we (the collective we) ended up with the abandoned places and people I witnessed from West Virginia to outside Gyumri, Armenia to along the rail line from Tbilisi to Batumi in the Republic of Georgia, the United States and the world were awash in discussion of banks "too big to fail" and of financial bail-outs of big financiers and even of whole countries. The field of STS introduced me to the critiques of corporate science and of "Big Science," and my husband, who had been a sustainable farmer in the 1990s before organic was available at the local Walmart, pointed me in the direction of understanding more about "Big Ag."

Reading the research and terms of other scholars, I realized that the abandoned places or places being abandoned "out in the country" that I witnessed in the United States and abroad usually were outside "commutersheds"—that is, they most likely were not going to become bedroom communities for growing metropolises, or, like that waterfront in Brooklyn, pivoted toward leisure and luxury. Nope. They were unmoored. Untethered by? By what? Who had let them get this way? In such different places so far away from each other? How could this be a communist, postcommunist, postindustrial, deindustrial, industrial, rural, *and* capitalist twentieth and twenty-first phenomenon?

Who had let the rural industrial places *fail*?

And, had these rural industrial spaces really failed when rural places, still, provide most of the natural resources of the material goods of our lives?

Why wasn't this fact seen? Appreciated? Why weren't the people tossed aside when they were no longer needed accounted for? Nurtured into something else? These spaces acknowledged? The reduction in the need for labor to extract these natural resources planned for?

Big Banks. Big Ag. Big Science. Even countries. They had and were all failing, and, the centers of power in the West committed BIG resources to propping them up, restructuring them, forging new entities, extending lifelines, replacing their extractive roles (Big Science) with potentially renewable ones (clean energy).

I repurposed the term "Big" as a way to think about the role of the beyond-the-commutershed rural places that had/have become largely single economic sectors over the last hundred or so years (with a nod to the longer colonial legacy of plantations, "banana" republics, etc.), these "big" "rural" places—hidden from discussion or view from society's political and power centers, and, so ubiquitous as to be like the air we breathe, a fact of life available seemingly to all until something makes it so polluted it cannot be used by anyone.

Big Rural—rural spaces connected across the planet in how they support the tangible, touchable needs and desires of, and the energy to fuel, modern life.

I examine the role of science and technology research and practice in constructing a specific case connected to this Big Rural, the rural energy production industrial space of the Pocahontas Coalfield. I examine what this large technical intervention has contributed socially and economically in this rural industrial space, and then propose how science and technology, as academic practices and research and development, ought to proceed in this rural industrial space or Big Rural more generally. I propose gaps in the academic fields my examination draws from, providing additional context for constructs such as "rural" and "industrial" and "Appalachia." I then provide essential background on why the Pocahontas Coalfield specifically merits an examination as a type in the Big Rural.

In my review of texts in rural sociology, rural studies, urban studies, and Appalachian Studies I have yet to encounter abundant analysis and engagement regarding the science and technology that serves as the base for the coalfields of this region. However, in STS, through the lens of the history of technology, Anthony Wallace's detailed history of the coal mining town of St. Clair, Pennsylvania stands as an exception in its discussion of the geology enabling, then disproving the capacity for economically viable mining in that location.[1] The threads of potential current inquiry into the science and engineering enabling the Pocahontas Coalfield continue to intrigue me; even a casual flip through *Mining Magazine* reveals connections to their continued construction of this place not only as a material space but also in material culture. These bring me to lines of inquiry such as:

- What science and/or technology enables a rural industrial space? How does one delimited industrial space such as the Pocahontas Coalfield fit into the theoretical concept of Big Rural?
- Who performs the Pocahontas Coalfield's science and engineering?
- What technology constructs the Pocahontas Coalfield?

- What kind of money follows these scientists and engineers? Who employs them? Where? How?

STS affords me the opportunity to layer in another question, absent in the discussion of this region: *How ought scientists and engineers engage with this space they also construct?*

Numerous other scholars have addressed the issues of labor, identity, power, region, culture, politics, gender, race, ethnicity, and capital in the case study subregion in question.[2] I reference their work when appropriate, but as I stated earlier, my concerns pull from narratives less obvious in most popular or academic discussions of the rural: LTS, laboratories, philosophy of technology, philosophy of science, sociology and history of technology, and urban studies.[3]

I also examine the Pocahontas Coalfield as a technological intervention performing economic stimulus, creating commerce and the economy.[4] Far from a new topic, Appalachian scholar Talmage Stanley completed a thorough, though not without issue, examination of the "Poca Fields" [Pocahontas Coalfield] in both his dissertation and in an academic text. For family and personal reasons, Stanley was drawn to review the role of the Pocahontas Coalfield mid-twentieth century in fulfilling its promise of middle-class prosperity. He outlines important background statistics for this industrial space:

1860: McDowell County [site of the West Virginia Pocahontas Coalfield] stood at 1,535 persons.
1880 = 3,074; 1890 = 7,300. [First industrial scale mine opened there in 1883.][5]
1900 = 18,747.
1940 = 94,354 persons. The coal boom associated with World War II brought the peak to 100,000 persons.[6]

Stanley displays a table tracing Carroll and Pulaski Counties in Virginia (which drew workers into McDowell to live and to work) and compares them populations for McDowell County from 1860 to 1940. While Carroll and Pulaski grow from 8,012 and 5,416 to 25,905 and

22,767 respectively (a three to four multiple), McDowell County grows from 1,535 to 94,354 (a multiple of sixty-one), the latter figure not taking into account commuters from counties adjacent to McDowell.

Stanley draws the links from this coalfield, its meaning and promise to local people, and its place in the global economy, and it as a kind of space in what I term "Big Rural":

> the heavily industrialized and urbanized Pocahontas Coalfield was a place of extraordinary wealth, power, promise, and possibility. In many people's lives,

> the Pocahontas field came to be associated with the struggle to lay claim to a future in the American middle class as it was defined within the social context of that place. . . . At the same time these processes were changing, they have also been continuous with the history in which the coal-rich Cumberland-Allegheny Mountains were a part of global patterns of consumption and accumulation. . . . The opening of the Norfolk and Western Railroad west to markets across the Ohio and east to shipping terminals at Tidewater ports propelled a history of explosive expansion and investment. In order to acquire and retain absolute control over coal and the certain profits to be made from it, capitalists undertook a frenzied reorganization of places in Appalachia, the South, and even the world. . . . The Pocahontas Coal Operators Association retained Phil Conley as its propagandist to use letters, pamphlets, magazines, speeches, articles, and books, to exhort the public about the benefits of Pocahontas coal. Conley constantly brought attention to the benevolence of coal companies, the progress wrought by the coal industry, and the benefits to civilization from burning the coal mined in . . . [the Pocahontas Coalfield].[7]

To the discussion in the *Preface* of pastoral Appalachian purity and degradation, this excerpt layers in industrialism, class, population growth for coal work, aspiration, capital, capitalists, and coal operator propaganda. These lay significant and important groundwork for parsing the Pocahontas Coalfield as a technological and scientific space, and, as an exemplification of rural industrial spaces more generally.

In *Big Rural*, I examine the role of science and technology research in constructing the Pocahontas Coalfield, what this science and technology has contributed, the throughline from current science and technology, and then propose how science and technology, as academic practices and research and development, ought to proceed in this rural industrial space or the rural more generally. I then link these to a reconceptualization of the rural in its relationships to power, centers of power, and democracy, and push for policy reconsideration to restructure "Big Rural" as other "too big to fail" entities have been restructured, redesigned, reconsidered, and/or invested in financially.

Within the American context, one move to discuss technology in terms of the rural would be to focus on the potential impacts of specific technologies: for example, the potential of broadband to help rural artisans reach broader markets, or by contrast, the potential for broadband as a force of potential disruption of some kinds of imaginary pristine or homogeneous rural culture. Another move may be to analyze the hypermasculinity of large rural machines such as industrial combines or draglines and how those dictate how and what serves as masculine or feminine and explain this phenomenon for the benefit of an academic audience outside of the countryside. Yet, another move could be to reconstruct the cultural changes stemming from the introduction of a wide range of machines, from new sitting hair dryers at the beauty shop (in places where people still go to "beauty shops") that emphasize a technology most likely to be used in a form for which it was designed. Likewise, I could examine the latest military style video games on rural children's lives (where rural inhabitants are more likely to enroll in military service) or cellphones for keeping track of medical issues of rural health or other kinds of infrastructural technology such as a lack of sewage infrastructure on rural areas in Appalachia or on Indian Reservations and how that kind of technological system (or the lack of it) impacts rural health and the environment.[8]

In theorizing and addressing the technology of the rural, I opt for a very different direction. Rather than focus hyper-locally on specific machines in the rural, or at work in the rural, I globalize experience but do not flatten or erase it or make invisible any one kind of inhabitant or geography. Instead, I seek to make the case for these interconnections that stretch beyond a hyper-specific rural space. Again, the field of STS offers powerful analytical tools such as LTS, laboratories in the field, coproduction, problems of categorization and classification, and issues of science in the polity for theorizing the rural industrial space.[9]

Rurality has not been a traditional object of inquiry in STS and this work fills a gap in STS theoretical examination and exploration, in the creation of LTS, how rural people in the rural industrial space function in LTS, and the results and risks of large-scale technological interventions. The last fifty years have seen a plethora of scholarship on this space and the greater Appalachia region incorporating deeply personal reckonings, labor war witnessing, and environmental wreckage testimonies—an often intimate scholarship. No on-the-nose answer exists as to what renders the Pocahontas Coalfield a highly technical space, but to circle closer to a range of answers, I draw from what I have thus far encountered in academic scholarship and contend with technologies and the creation of rational lands and spaces through STS.[10]

In popular culture, in social media, from the executive branch, from politicians and company men, but also from scholars—I am weary of tired rhetorics and of myths and monsters: the easy and comfortable way from any "side" of the current political currents—whether it be the coal industry characterizing President Obama as an evil job killer, or environmentalists protesting surface mines on site antagonizing coal miners when the real bosses who call the shots live hundreds of miles away. As a scholar, I do not bring any new ammunition to these entrenched fights. I also generally have been far more interested in the heart of darkness in us all or, maybe said differently, but for the grace of God there go I. I do not think fueling fires or poisoning wells gains me further insights in terms of understanding the snow globe that was and is the Pocahontas Coalfield, and, how, moving forward, that industry now forever mise-en-scène can be switched up.

I am also fatigued by the discussion of abstracts in action—capital acts, capital does, rather than naming capitalist names. While I draw upon the work of David Harvey and others who work also in general constructs such as "the economy" and "community" or even, "rural"—in the case especially of the latter, I hone down the set of definitions. Though I contend with working from a broader view of my subjects at hand and search for a type, a representational quality, and how they function as examples, I also want to see all the subjects involved as human: if I afford myself the stunted intellectual road of painting concrete or abstract enemies, spin academic yarns of capitalist or backward barbarians, then I open the door for any numbers of targets for whom I have deep sympathy to be also rendered subpar, to be likewise abstracted or "othered": the poorest folks, the lowest caste, the drug users, the left-behind elderly—those in deep struggle living in the Pocahontas Coalfield region. This space's cyclical booms and busts have transferred the fountain of luck so often that each subsequent pour has left new sets of folks high and dry.

THE IMPORTANCE OF FIELD OBSERVATIONS: *ACADEMIC AND STAKEHOLDER CONVERSATIONS THAT NEED TO HAPPEN*

From April 2015 to September 2016, working with mining and oil engineers, I read more than seventy-five papers in mining safety engineering, reclamation science, fish and wildlife, economics, hydrology, mining engineering, epidemiology, and biological systems engineering. I interacted directly with more than forty researchers in these fields. I kept abreast of research and economic trends and issues in coal and natural gas production in the United States. I visited major Appalachian coalfields as well as natural gas shale plays. I attended a conference on land reclamation and one on society and energy production (from a fossil fuel perspective). I listened to mining engineering colleagues discuss climate change, energy abundance, energy independence, energy policy, Environmental Protection Agency regulations, and jobs. At these two scientific conferences, I ran into no recognizable participants affiliated with Appalachian Studies or STS. Likewise, in 2017, at the Appalachian Studies Conference, I ran into no academics or colleagues from the cadre of more than ninety scholars, scientists, and engineers I worked with from seven universities in Appalachia.

Though I contend with the facts on the ground and the results of the single sector coal mining industry in the Pocahontas Coalfield, my aim in macroanalysis remains *systems* rather than individual or even specific rural industrial spaces. In no work that I reviewed that also engaged heavily with the social and cultural context, save for Anthony Wallace's (STS) and somewhat in Jerry Bruce Thomas' (Appalachian Studies) work, was the black box of scientific and technical practice in the Appalachian Coalfields cracked open. With my hands in STS, one foot in Appalachian/Rural Studies/rural sociology, one foot in the cross-section of the scientific and technical practice of the fossil energy industry in Appalachia, and, also attending a range of nonprofit and economic conferences aimed at Appalachian "transition" away from fossil energy sector read *coal* dependence, I remain struck by how definitively and completely these latter three sets of major stakeholders in Appalachia operate with little to no cross conversation. When I spoke with professionals engaged in the fossil energy sector directly or as scientific and technical experts, by far and away the most impactful cultural group on Appalachia itself, their perspective and outlook on Appalachia, their vision for its future, their justifications for its social and economic issues, remain largely unexamined in STS and other relevant literatures. In consideration of concerns for a conflict of interest, while working with an energy group, I could not also then take on a new set of subjects for qualitative interviews beyond the grasstops I engaged. Still, my work within the energy sector convinced me of the necessity of working for inclusion of consideration of the role of current scientific and technical expertise and the culture of that expertise in the rural industrial space, in specific spaces, and then also in Big Rural.

Again, major gaps remain in understanding the cultural practice of scientists and engineers in fossil energy production in the Appalachian rural industrial space. I contend with this need by emphasizing the saliency of science and technological research practice in constructing the rural industrial space and by underscoring that array of practices. While these scientific and technical practices and their accompanying culture or use may seem obvious to practitioners of STS, not every other reader who may encounter this work will necessarily be as prepared to accept that premise. Had I not worked in the fossil energy sector and read as widely as I did while engaged with it, I too would have presented a less nuanced examination of the actors behind the region's "capital" and its "technology."

One aside worth recounting: in early 2016, at a conference of major regional and federal economic development funders, the question arose of whether it is safe to farm in formerly coal mining regions (though coal is not the only mineral mined in Appalachia or with potential impact on agriculture). For economic development work I engaged in adjacent to research, and, at that time, for my professional work with the Virginia Center for Coal and Energy Research, I had reached out to a range of scientists on this exact question. I raised my hand and answered the inquiry. I followed up by emailing several leaders from this group. I sent them the contact information of the scientists to engage. I interviewed scientists, created a set of protocols for interested parties to follow, and published them on the web.

The very people most needed to answer this inquiry with expertise so that sound economic development decisions could be made were not only not in that room, but from my field observations, operating in an entirely different social and professional circle in the region. From participant observation working with scientific and technical researchers, from my literature review of Appalachian, STS, and rural sociology and studies, and my analysis of conference materials and nonprofit conferences, here is what I have surmised.

The public relations rhetoric of land fights regarding surface mining and natural gas pipelines has deepened an already large communication rift between those parties, academic and nonprofit, seeking to transition away from a fossil fuel dependent regional energy economy, and the scientists and engineers that work for those fossil fuel industries.[11]

The cultural studies and nonprofit community may not grasp the need for understanding the current scientific and technical practices of the current rural industrial space—the tropes of nonindustrial Appalachian experience can mask the current and historical contaminations in a rural brownfield site that "seems" restored, where the scientific fact remains that each site, due to its unique geology, hydrology, and scientific and technical history, requires individualized scientific and technical protocols, and, could for generations.[12]

Because the rural space at hand also has a long industrial history, a site-specific technical history along with current technical advisement must be included and implemented in new plans for site intervention. If not, contamination, geological (literal landform or potential for landform shifts), hydrological (water moves and new issues have the potential to arise or resurface), and legal risks (to workers, to consumers) may not be mitigated.

To grasp the current on-the-ground economic situation, academics and nonprofits investing in economic and social change ought to track and follow the region's fossil fuels and other natural resource-based industries and their accompanying partner organizations, academic departments, lobbying efforts, conferences, federal agencies, PR campaigns, donations, and so on. My experience as a participant–observer convinced me of the need for bridges and engagement to be built across this wide aisle.

If one takes away nothing from *Big Rural*, then take this: in terms of creating regional culture and constructing regional rural space, the fossil fuel and other natural resource extraction industries continue to punch above their weight, and, to reiterate, they use science and engineering to accomplish economic and technical needs or wants, and, as an industry, always have. Those scientists and engineers receive their professional training at the region's research universities. There are large stakes for those people and the departments in which they are educated should local extractive industries shift. The employ of these local scientists and engineers and the departments where they trained form part of a system of support for the local fossil fuel industry.

For example, the energy sector has a range of stakeholders and coproducers in Appalachia coproducing the "energy sector" together. Reducing this web of contacts and stakeholders only to capital or owners makes for tidy but questionable assertions. The human actors with power in the fossil energy sector in the Appalachian Region, at minimum, include: citizens, vendors, scientists, engineers, companies, advocacy groups, lobbyists, economists, politicians, federal government agencies, media, marketing firms, educational institutions, and so on.[13] Though I hesitate to situate science and technology research as underlying or as the common throughline among these actors, without adequate scientific and technical expertise the energy sector could neither meet economic imperatives nor regulatory demands.[14]

We could approach this web of relationships through several lenses. Science and technology research for the energy sector coproduces with these actors: all these stakeholders constitute "mining engineering" working in tandem, with the more closely related actors spinning together while "investment capital" buoys the spins.[15]

I could also recreate this web as a chart to examine the rural industrial space as a kind of system with: citizens, government, scientists, engineers, corporations, vendors, corporate advocates, economists, lobbyists, media, politicians, educators, public relations, marketing, floating on a sea of capital from investors.

I pull the investors to the side as though they may underwrite the rural industrial space, they rarely engage in the place of the rural industrial; their money funds it, but they do not engage in place creation in the here and now.[16] I demarcate "citizens" as those rural industrial space participants affected by what political scientist Deborah Stone terms "corporate actors" meaning entities that can exhibit so much control over individuals that the power of these entities ought to be especially limited.[17]

In my areas of theoretical review, I encountered abundant analysis of citizen, politician, government, and media stances or participation in the creation of the rural. I encountered corporations and investors black-boxed under "capital." I have encountered very few study areas (Science and Technology, Appalachian, Rural, economic geography) concerned with theorizing the science and technology actors in the rural industrial space, especially at the macrolevel.

While working for the energy group at Virginia Tech, I met a number of people from my home state of West Virginia as well as from Southeastern Ohio, Appalachian Kentucky, and Pennsylvania, for whom some kind of affiliation with fossil fuel energy had provided a good living, and who earnestly believed the tradeoffs for this fossil fuel's deployment were worthwhile, and could cite solid statistics on what cheap fossil fuel energy had created in the United States in terms of material prosperity.[18] These were people who largely went to college locally—or not far from home—bright people with a proclivity for math, technical modeling, or a love of geology, hydrology, and statistics. I met any number of women involved in geology, forestry, fish and wildlife, and reclamation science—people who belonged to groups like one, which networked in smaller energy-focused cities: Pittsburgh, Pennsylvania; Charleston, West Virginia; Lexington, Kentucky. I contend that the scientists, medical researchers, and engineers with whom I discussed the economics of the region, and its poverty and corruption, do not conspiratorially contribute to either of the latter. In 2015 I argued with a prominent regional journalist when he insisted that these scientists and engineers indeed sought to take the side of the fossil energy corporations; I asked him if he understood how scientific careers worked, the need for peer review, the unlikely obvious or explicit for hire by industry of the university-based researcher. I countered, as unlikely as it may seem, that university-associated scholars may stand in a different worldview, or even paradigm, with respect to social, economic, and scientific facts regarding the region. I am not seeking to apologize to or for anyone, but to understand how this fossil energy sector constructs a rural industrial space, and, if we know this, whether we can intervene with other science, research, technology, means, and people for outcomes other than those which we currently have.

I can see the hackles up on some readers' backs: but, but, but—you have folks here like a *Don Blankenship*. Read on—I theorize and contend with Massey Energy, but, again it is not that I do not have a dog in this fight: I am the dog, I am of the dogs. And, contrary to popular belief, we can learn new tricks. Yet to mix metaphors, we first also must understand in some new ways the bird in the hand.

In STS we have a tradition of examining laboratories.[19] To put it plainly, the coalfield is a large laboratory; technology and science have made field laboratories of every Big Rural space: the millions of acres of industrial agriculture, the hundreds of thousands of acres of mines, oilfields, gas wells, pine for pellet harvesting, fish farms. Big Rural spaces like the Pocahontas Coalfield are created by scientists and engineers: scientific and technical workers in (insert an applied science!) in geology, hydrology, mining engineering, safety engineering, reclamation science, materials engineering, and so forth. They turn the "natural" environment into a cauldron of varying and various chemical inputs—legal and illegal, hydrologic machinations, geologic dumpings, and externalization of environmental costs shaftings.

In the case of the Pocahontas Coalfield, from the latter quarter of the nineteenth century, the whole "built" environment, from the employ of people to the railroad to the movie theater to the churches, was constructed as an extension of people as technology to service the coal industry: nature and people as nature as an industrial laboratory. Nature and people as technology, serving up coal. Coal as the institutional force—the power of the king; the regional panopticon—King Coal sees you everywhere you go; even if you don't work for the king, you rally to his defense.

Without science and engineering, there would be no Pocahontas Coalfield or other Big Rural caudrons cum laboratories. I question the aim of these sciences, their morality and ethics, and their scientific knowledge creation.

For ethicist and theorist Dorothy Nelkin information creation and access remains political, including scientific information, and, worthy of state analysis, citizen scrutiny, and effective policy. Likewise political, the diffuse nature of some technical and scientific-created or addressed issues makes a cohesive policy attack through collective citizen response weak or ineffective.[20] My analysis, and in the context of recent coalfield organizing, is that while marching on a specific mountaintop removal site can be orchestrated—we can gather people together and go on a march, real change in certain rural industrial spaces will come only through shifting America's energy mix, as reorganizing or repurposing coal-fired (or natural gas) power plants takes top-down, federal or at least, powerful and willing, state will. The latter may be what we sorely need, but the former (marching) stands as more likely to happen in terms of tactical orchestration and achievement.

With respect to diffuse issues and specific policy and state-making, as one coal boss asked me (and I paraphrase): Where are all the protests against the same massive earth-moving machines being used to expand an interstate near Boone, North Carolina or to make way for yet another big box retail complex with cheap stuff from China where hundreds of acres are paved over and not one damn person monitors the creek change from the parking lot runoff? I note his detraction from coal industry policy and environmental monitoring, but he addresses a valid observation and makes a reasonable oranges and oranges comparison. My point again: he is not irrational. Yet, the sane policy response is to regulate both coalfields and big box interventions.

The Pocahontas Coalfield space extends out, far beyond, and encompasses and touches, places well beyond its borders.[21] In STS, we examine categories and classifications.[22] Can I trace every connection to this coalfield? Of course not. Yet, the science and technology enabling the snow globe cum laboratory cum socio-techno space is, largely, loosely identifiable as are the companies and people of the Pocahontas Coalfield's "capital" and how these first groups and these second groups connect, overlap, diverge, converge. Furthermore, government agencies, research agenda/s, and support research and development (R&D), also shape these fields. Then there are the railroad companies, the transportation engineers, the logisticians, and the financiers. A complete history of this space would be unwieldy, and again, I do not aim to illustrate history, but to illuminate potential spokes out and in such as academic research and development, investment, advocates constructing the Pocahontas Coalfield. Much scholarship exists on the miners, the more local coal barons, African American communities, immigrant communities, and the prior inhabitants displaced.[23] I seek the science and scientists in, and engineers constructing, this kind of space.

SPOKES IN AND OUT OF THE POCAHONTAS COALFIELD

In his description of scientific norms mid-twentieth century, sociologist Robert Merton raises the questions of science for public benefit and describes the acquisition of patents by scientists such as Albert Einstein in order to ensure their work remains available for public use. Merton highlights that by contrast, "scientists have been encouraged to become promoters of economic enterprise."[24] That the focus of mining engineering departments rests on working hand in hand with the mining industry comes as no surprise. I assert that the institutionalization of their relationship along with the institutionalization of mining engineering itself can contribute to the codification and ossification of scientific expectations for the rural industrial space where this technical expertise prevails. Mining engineering contributes to the construction of mining in the rural mining space. As the main science intervening, a paradigm shift in the goals of geology, mining engineering, and even sciences such as forestry and wildlife, which assist mining companies with science protocols to meet regulatory environmental standards, would need to manifest.

Earlier in the *Introduction*, I made known my preference for tracing actors in the polity rather than continuing to discuss constructs such as "capital" as animated. In terms of constructing the Pocahontas Coalfield, the following four mining engineering programs illustrate the focus of this expertise for hire. I pose the need for new assessments of the rural industrial space with respect to the intervening science and technological expertise. I call for a civic science, a science for the public benefit, and a new reckoning of what "public benefit" may mean, and, by whom:

University of Kentucky—College of Engineering, Department of Mining Engineering
The Pennsylvania State University—College of Earth and Mineral Sciences
Virginia Polytechnic Institute and State University—Department of Mining and Minerals Engineering
West Virginia University, Department of Mining Engineering

A brief review of the details of some of the programs and promotions of these departments in appendix A reveals how openly and very typically the

Pocahontas Coalfield also is in construction and performance as a space of sophisticated scientific and technical intervention and of corporate science. In short, these departments promote the federal and corporate funding they have received, the sophistication of their laboratories, their multitude of corporate partners, the success of their alumni in securing high-paying jobs, and the academic publications of their faculty. What may seem odd would be how the role of mining engineering has been ignored in the cultural, economic, and social enactment of the Pocahontas Coalfield. This omission reveals gaps in the intellectual conceptualization of this kind of rural industrial space, which may then also contribute to policy, governmental, and social change gaps. To recall STS theorist Sheila Jasanoff—in the case of democracy in our country, we must include and link examination of science and technology.[25]

A perusal of the multitude of jobs that accompany science and technology in a coalfield space rounds out the many scientific and technical actors in the coalfield space beyond miners. See appendix B.

To these, a small sampling of jobs from the American Society of Mining and Reclamation completes our glimpse into contemporary university-led scientific and technological practice in a coalfield space[26]:

- Environmental Project Manager, HDR—Austin, TX, Posted Saturday, June 10, 2017
- FORESTER II, Boscawen, NH, Posted Saturday, June 10, 2017
- HYDROGEOLOGIST II, Concord, NH, Posted Saturday, June 10, 2017
- Forester I Haverhill, MA, Posted Saturday, June 10, 2017

I could continue and drown us in the number of employees at the U.S. Office of Surface Mining Reclamation and Enforcement as part of the Department of the Interior.[27] I could provide textual and visual cultural analyses of industry journals such as *Coal Age*, *Reclamation Matters*, and then critique the rhetoric employed on the website *Friends of Coal* and the machines on display in the expo of the Bluefield, West Virginia Coal Show. Swimming in details, running lines in and out, may offer us a repetitive performance of the technical expertise, the great many expertise threads in the web buoying the performance of the Pocahontas Coalfield.[28]

This aggregation of coal career and coal education artifacts becomes lost in the understanding of a rural industrial space when we draw our fields of academic inquiry too narrowly. If we do not search for the current science and technology buoying that space, we lose the threads of current *interests*—corpora-university, corpora-federal, corpora-political. By focusing on miners and "barons" we lose so many other spokes, so many other threads, so many stakeholders cum advocates—actors in this field. I can only liken it to reducing the full drama of Shakespeare to a mere two players.

Big Rural also aims to unveil several sets of other "hidden" actors in a specific rural single sector economy, to examine the role science and technology played in advocacy for the very earliest days of this single sector economy, the subsequent later role of automation in the diminishment of this single sector as a major employer, and to examine policy, institutions, and models for contending with the remnants and repercussions of this sector and its techno-cultural footprint in this rural space, and, as a model to rethink Big Rural structurally.

Though the qualitative case study and the corroborating quantitative data sets I employ focus on the central Appalachian coalfields primarily of Southern West Virginia and Southwest Virginia, my attending analysis includes in its perspective these same issues in rural single sector areas more broadly.

AN ASIDE INTO THE "RURAL"

Rural must be situated culturally and historically. I dive into the terms of rural space as it stands as a political and municipal entity. Understanding how we arrive at what has become a default and unquestioned conceptualization of the U.S. rural sets the stage for how a kind of rural both unifies across many rural and ruralities and subverts the culturally de facto hegemony of rural in the United States as White and male.

Rural can be synonymous in concept with how we currently consider and use the term "offshore" and vice versa regarding where rural has been located and what its function has been historically in the United States. A brief detour into history brings us to the late Middle Ages and early Renaissance when Europeans were searching for riches beyond their borders. Once located and pillaged, they exported these raw materials back to Europe, thereby eliminating the need for accessing certain raw materials in Europe. With time, colonists in North America seized raw materials, conquered locals, and built cities. The cities needed raw materials and colonists stopped exporting all that was of use to Europe. As demands for raw materials grew internally, colonists took more territory to satisfy these demands.

Now, with the specialization of production of raw materials in different U.S. sectors, rather than producing raw materials for all the needs of personal or corporate existence, the raw material production or extraction is Not In My Backyard (NIMBY) or offshored, with cheap energy and cheap transportation fueling the further offshoring to the cheapest producer of anyone's personal or corporate needs. In a city, I would produce no raw materials and source all my material products from off-shore. In the single sector rural, I may extract or grow a raw material, but I do not produce the machines by which it is utilized, and, in most cases, all other things I need to survive come from what is offshore to me.[29]

Certain kinds of internal U.S. offshore or rural bubble up in larger cultural, pop, or policy discussions, and others remain obscured. Now and again in pop culture, we see, for example: White cowboys working hard in producing cattle for our consumption. White Midwestern male farmers giving us the grain for our daily bread. Arguments breaking out about whether it is okay to call a White, mountain dweller a "hillbilly." White American presidents and presidential contenders sporting rural gear—cowboy hats, flannel shirts, baseball caps—all to remind us of the original hard-working proto-American: the White, rural *Dirty Jobs* male.[30]

In the larger cultural hegemony, we do not associate Black cotton producers or pine plantation workers in South Carolina as rural. We skip over Indian reservations and indigenous spaces as rural. We ignore Latinx migrant workers harvesting onions in Georgia or grapes in California as rural. We do not include Asian or Latinx or African meat processor floor workers in Iowa or North Carolina as rural.[31]

In fact, with respect to Native North Americans, if we reside in the rural East, we ignore how we came to own land in the rural East; or, even if poor and White in the rural East, we have developed myths of origin for our places on that land, whether we are "native" "Appalachian" (which does not indicate indigenous) or culturally "Southern" waving a Confederate flag to honor what, in human time, is a very short "heritage," or potlucking in the upper Midwest as a real Iron Ranger (White) or celebrating our French roots in New Hampshire or Louisiana, being "born on the Bayou" and so on we go with our internal myths of rural, likewise with our designations as *native* New Yorker, Angeleno, Chicagoan, and so on. The short histories of any White U.S. heritages make for surprisingly long memories. In this respect, we do not differ that much from other White colonial conquering myth makers, such as the Afrikaners, with their and our myths of land rights, ancestral land claims, biblical claims, and so forth.[32]

Both these rural—the offshore and the myth—work in the background, usually invisible in larger political discussions of the rural in the United States. Unlike India, which has national discussions regarding its rural populations and also national rural policies, for the centers of U.S. society—its cities—the rural, especially the rural that is not White and male, largely remains invisible to those in power, Big Rural being so necessary in terms of material production as to be rendered invisible.[33]

Another layer to defining the rural comes in how theorists grapple with political terms for what lies outside metropolises. Turning again to STS and its work on boundaries, or what makes one object fit into one category while another fits into another (here I give one nod to structural linguistics and understand that not only STS is concerned with object boundaries), the demarcations I prefer also delimit the narrative I produce.[34] In my demarcations, I institutionalize and thereby rationalize (make rational and controllable) what may otherwise be uncontrollable or not truly able to be delineated. For example, although as a scholar I have come to associate constructs such as rural and urban as on a continuum, especially in late capitalism, not as opposites, I still draw permeable lines around each in order to contend theoretically with their relationships on this continuum. I liken this accounting to the territory in translation where one language does not have an exact cultural representation or practice in its language signifiable also by only one word in another. Classification and more accurate or "better" naming may not actually erase that gap, but they allow us to move forward with other meaning making..

By tracing how my vocabulary shifted as I discovered terms new to me, and struggled to make material what I needed to contend with theoretically, I intend to complicate rural as a construct. Rather than trace each term in my order of discovery, I explain terms and their employ. In so doing, I offer unfamiliar readers this considered set of terms for their own use.

Micropolitan, used by the U.S. Census and by urban planners, denotes an area without the population of a large *metropolitan* area and which is still more connected to a rural or less populated area. Of course, it, like the term metropolitan, is not a fixed term. Though smaller in population, a micropolitan still may have many of the features of a much larger city such as cultural events, crime, White flight, and so on.[35]

Metropolitan, with its flexibility to indicate a city and a city-dweller, became for me a place of contrast for both micropolitan and rural. The micropolitan could only be outside a metropolitan's other spheres—suburbia, exurbia, and beyond a metropolitan *commutershed*. A micropolitan was unlikely to have an exurbia, but it could also have a suburbia and its own commutershed.[36]

Rural as geography, and as opposed to nature, became that kind of place beyond the micropolitan and the metropolitan, and usually beyond their commutersheds. Being beyond a commutershed depends on the culture of work-commuting in a region. Commuting two or more hours one way by rail to work in New York City is not unheard of, yet, that same amount of driving to work one way to another metropolis would be impractical or highly undesirable.

A *metropolis* comes close to another useful term, the metropole. The *metropole*, as a collective, consists of the population centers also at a society's center of commerce, government, industry, and so forth. Offshores send their material goods to metropoles.

Over the course of time, the metropole can also shift. In the case of the United States, its White inhabitants initially sent raw goods to Europe, their (the White inhabitants') metropole. Priorities shifted and the White Americans chose to keep some of the raw goods in their "own" territory as their own metropolises developed. Over time the United States shifted from a largely rural population trading to metropoles internally and abroad, to an industrial rural supplying its own metropoles or others and increasing imports from other regions around the world.

Each *industrialized* society may have its own version of the rural and the metropole. A flow of goods characterizes these two constructs in the context of this work. Certainly, goods now flow around the world with what seems an impossible array of origins. However, this flow can be traced, largely as part of Big Rural, and this flow also tells a story of the function of a kind of rurality. This flow seems ubiquitous and immutable (Big Rural), and to the metropole consumer, and depending on the product, also to a micropolitan and rural consumer, invisible.

This invisibility only breaks down partially depending upon the local industry. For a miner, the source of cheap energy in a hinterland[37] mine may be obvious, yet he (as they are most often he's) thinks little about the sources of his cheap food. The monologic farmer profiled by Campbell, Bell, and Finney in *Country Boys* may fret about the cost of fuel for his massive farm equipment, but he spends little time considering the impact of the cheap energy on its site of extraction, and so the list goes on.

Raw material production areas, professions, geographies, and so on in the United States usually are not only located well outside most metropolises but also away from each other. Their commonalities in experience, concerns, interests (in the Marxian sense), problems, and so forth are little linked, especially theoretically or politically. STS brings a concern with networks and invisibility, seemingly undeterrable systems, to the fore and to focus.[38] This invisible system of *industrial rural*, offshored internally, links up to rural, raw material producers externally, making a kind of rurality (Big Rural) the backbone of *big* industrial, *big* cities, and so on, enabling those people in society's centers to be concerned with pursuits not directly related to growing food or creating or extracting other raw materials, and, in turn, enabling various industrial rural to hyper-specialize.

Whereas a U.S. homesteading family in the nineteenth century may have had all members farming, selling off surplus, or producing surplus in a variety of goods for external consumption, especially in the U.S. Northeast, by contrast, the flat land and rolling hill land and/or more accessible areas south of the Mason-Dixon line were already a kind of industrial rural, concerned primarily with raw material production for export to the "industrial," read *factory-laden*, North.[39] For example, cotton production in the U.S. South occurred under totalitarian circumstances, either by enslaved Black laborers or later by sharecroppers, Black and White. That raw material production was industrial in scale, with enslaved laborers as the principal enabling "technology" and enslavement as the enabling industrial institution.

AN ASIDE INTO APPALACHIA

If current scientists and engineers intervene in the Pocahontas Coalfield as a scientific, technical, and technological space, then why are scientists and engineers not the major subjects of much Regional or Appalachian scholarship, social change practice, or cultural examination? By contrast, the Pocahontas Coalfield sits in Appalachia, whose many competing and overlapping cultural associations or perceptions have attached themselves to the Pocahontas Coalfield as much as to other sections of the area, with or without basis. In the larger U.S. culture, the perceptions of Appalachia birthed by local color fiction in late nineteenth-century popular national magazines struck root.[40] Serial pieces on lost "mountain ancestors," spun tales of salt-of-the-earth, clannish, feuding, poor, White, uneducated, male, moonshining hillbillies.[41] This prototype demonstrates remarkable persistence, mutating and adapting, such as current preconceptions of Appalachians as poor drug users, or with rotten teeth from "Mountain Dew" mouth. The peoples in West Virginia, Kentucky, Virginia, Tennessee, and North Carolina Appalachia particularly continue to bear versions of these earlier fictionalized associations.[42] Changing with the times from the *Beverly Hillbillies* to the *Dukes of Hazard* to Jesco White to *Buckwild* to the *16 and Pregnant* television series to the more recent fascination and fury with Appalachia as "Trump Country," these many iterations still largely dictate to the wider U.S. imagination what supposedly *really* ranks as Appalachian: a mountain-bred White coal miner or White mountain man. Or, his kin. In every case, these people are found to be in deficit: in the 1960s, they were too poor and thin,[43] now, they are poor, fat, and in ill health from drug use, diabetes, heart disease, black lung, or other chronic illnesses.[44] They smoke or use smokeless tobacco; the men, when they work, do manual jobs and eschew safety gear or protection.[45]

Much like other projects of nationalism, literary fiction and scholarship about or even in the region have also helped define the region.[46] For example, in the case of the critical regional approach to Appalachian Studies, its scholars, writers, policymakers, regionalists, historians, culturalists, and so on also construct Appalachia. They contribute to the creation of the cultural production of the region known as Appalachia.[47] Social scientists proposed that a culture of poverty caused Appalachia's decadence and fatalism—these Appalachian people needed to be taught better, and how to adapt to modern society. Scholars focused on Appalachia advocated a version of Appalachia caught in exploitation. From the 1970s the discussion moved from Appalachia as an internal American colony, much like the kind of third world colony exploited for first world gains in natural resources or agricultural or mining production, to a world systems analysis approach, with Appalachia a peripheral space within a core economy.[48] Recent analyzes from journalists such as Chris Hedges to scholars such as Betsy Taylor parallel this core and periphery concept with analyzes of extractive Appalachia as a "sacrifice zone"—a space exploited so that more metropole or central places may prosper.[49]

Though Jerry Bruce Thomas in his book *An Appalachian Reawakening: West Virginia and the Perils of the New Machine Age*, 1945–1972 catalogs mining machinery and its effects on job loss, little of the sociological or cultural critique of the region engages or discusses science, engineering, and investment—all essential elements to wrangling and maintaining, constructing, the Pocahontas Coalfield into existence.[50] Thomas discusses how, since the 1950s, people in Appalachia missed out on the economic prosperity of the greater American experience largely due to drops in demand for coal and small scale agricultural products, the results of technological shifts in coal mining in Appalachia and farming in America's West and Midwest toward utilizing bigger machines with fewer people. Nevertheless, he does not address at any length one of the supposed reasons for U.S. economic expansion elsewhere during this time: scientific and engineering "progress."[51]

Though automation was a fear of many Americans by 1960 in terms of well-paying jobs being automated away (according to Thomas), during the 1950s and 1960s the U.S. government and industry touted much about the supposed progress of U.S. science and technology, especially with respect to its major competitor, the Soviet Union, and as a major driver of the U.S. economy.[52] Though current popular consumer technologies also appeared and were in homes across Appalachia during this time, like the economic situation in the Pocahontas Coalfield being single sector and extractive, industrial science and technology in Appalachia also focused primarily on issues of resource extraction: natural gas, oil, timber, coal, steel, other minerals such as silica and aluminum, and the Kanawha Valley's chemical industries.[53] In Appalachia, as elsewhere, these decades also saw the emergence of science contending with the environmental fallout of the industrial age such as the rise of reclamation science, environmental science, and so forth.[54]

CONSTRUCTING AN INDUSTRIAL SPACE

Scientific fields are human constructs as are the fruits of applied science. Rather than engage in an academic discussion on whether science has immutable laws existing beyond the human capacity to interpret them into the boundaries of fields of science, I draw from the STS tradition more concerned with the social choices and cultural choices of applying science. I deliberately use the verb "construct" as shorthand to draw from the sociological theory of knowledge "social constructivism." I do not dwell on this term's implications in language theory, but model why one version of Appalachia succeeds in popular culture while Appalachia as a highly technical space falters. I insert the implications for this gap in politics and power in the polity and policy.[55]

With respect to the term "industrial," I also reveal the construction of the Pocahontas Coalfield as an industrial space, replete with all the trappings of the industrial factory space: a move from hand production to machine production, new chemical and industrial processes, the importance of cheap energy and steam power, and the streamlining of work and workers, as well as the locating of these processes in a space that extends the factory doors to include the construction of civic space to service the needs of workers: housing, education, farming, food, stores, churches, civic groups, and so on.[56]

I outline gaps in the literature of STS, and Appalachian and Rural Studies, with respect to examinations of the role of scientists and engineers in the rural space. I define my use of constructs such as rural, industrial, Appalachia, and the term "construct." I propose science as a pursuit not only of corporate-driven science but also knowledge production for public benefit. I set the stage for my theoretical examination of the Pocahontas Coalfield as a rural industrial space constructed also by science and technology, scientists and engineers.

NOTES

1. Anthony F. C. Wallace, *Saint Clair: A Nineteenth Century Coal Town's Experience with a Disaster-Prone Industry* (New York: Random House, 1987).

2. For example: John Gaventa, *Power and Powerlessness: Quiescence and Rebellion in an Appalachian Valley* (Chicago: University of Illinois Press, 1982); Amy D. Clark and Nancy M. Hayward. *Talking Appalachian: Voice, Identity, and Community* (University Press of Kentucky, 2014); Ronald L. Lewis, *Black Coal Miners in America: Race, Class, and Community Conflict, 1780–1980* (University Press of Kentucky, 1987).

3. In urban studies can be found remnants of a once robust academic field: rural studies. By the 1960s many rural studies departments in the United States had shut down, leaving two prominent ones active: Ohio State University and McGill.

4. Douglas Reichert Powell in *Critical Regionalism: Connecting Politics and Culture in an American Landscape* illuminates his use of space over place in his discussion of various sites he examines in Appalachia (Chapel Hill: University of North Carolina Press, 2007). Drawing and extending from his use, while a place often merits

statutory description in legal codes, space connotes a place being produced, not fixed, and emanating beyond its statutory borders. While the Pocahontas Coalfield can be located on a map as a defined place, it is a space in flux in terms of what that space means, its purpose, its kind of citizenry, its technology and science. I do not employ the terms "space" and "place" interchangeably.

5. Stanley, Talmage A. "The Poco Field: Politics, Culture, and Place in Contemporary Appalachia." PhD diss, Emory University; 1996, ProQuest Dissertations Publishing. *The Poco Field: An American Story of Place* (Urbana and Chicago: University of Illinois Press. 2012).

6. Ibid.

7. Ibid.

8. For example: Robert LaRose, Jennifer Gregg, Sharon Stover, Joseph Straubhaar, and Serena Carpenter. "Closing the Rural Broadband Gap: Promoting Adoption of the Internet in Rural America." *Telecommunications Policy* 31, no. 6 (2007): 359–73; Roger Horowitz, *Boys and Their Toys: Masculinity, Class and Technology in America*, Hagley Perspectives on Business and Culture (New York: Routledge, 2001); Santosh Krishna, Suzanne Austin Boren, and E. Andrew Bates. "Healthcare via Cell Phones: A Systematic Review." *Telemedicine and E-Health* 15, no. 3 (April 2009): 231–40.

9. David Roochnik, "Socrates's Use of the Techne-Analogy," *Journal of the History of Philosophy* 24, no. 3 (1986): 295–310; Andrew Feenberg, "What Is Philosophy of Technology?" In *Defining Technological Literacy*, ed. J.R. Dakers (New York: Palgrave Macmillan, 2006), 5–16.

10. See Michael Hendryx and Melissa M. Ahern, "Relations Between Health Indicators and Residential Proximity to Coal Mining in West Virginia," *American Journal of Public Health*, (April 2008): 669–71. I also include epidemiological as recent academic cum political controversy regarding Appalachia. This space of the Pocahontas Coalfield circles on the debates regarding Michael J. Hendryx's and Melissa Ahern's work on correlations of health outcomes in surface mining adjacent living spaces. Their work opened up public health and epidemiological debate on why certain Appalachian spaces host so many chronic illnesses and whether these can definitively be proven to be work, and not culturally, related.

11. Jeff Turrentine. "Coal Is Literally Killing Us," NRDC, June 30, 2016, accessed on July 13, 2023, https://www.nrdc.org/onearth/coal-literally-killing-us; *Coal is Dirty*, n.d., http://www.coal-is-dirty.com; *Friends of Coal*, n.d., https://www.friendsofcoal.org/; Richard Martin, *Coal Wars: The Future of Energy and the Fate of the Planet* (New York: St. Martin's Press, 2015).

12. Christopher D. Barton, interview by Crystal Cook Marshall, July 2016. Chris Barton, Professor of Forest Hydrology and Watershed Management at the University of Kentucky raised these points during an author interview regarding contamination in mine regions and the suitability of coal mine regions for agriculture for human consumption.

13. As STS also has a theoretical tradition of assessing and including nonhuman actors [Bruno Latour. *Science in Action: How to Follow Scientists and Engineers through Society* (Cambridge, MA: Harvard University Press, 19870); Donna

Haraway, *Simians, Cyborgs and Women: The Reinvention of Nature* (New York: Routledge, 1991)], I have chosen to limit my theoretical discussion to those that are human. However, Actor Network Theory could readily be applied to the life-forms mimicking human habitability in the coal producing zones: micro-invertebrates, salamanders, dace, *E. coli* bacteria. For example: Nicholas Cook, Leigh-Ann Krometis. and Emily Sarver, "Inventory of Bacterial Impairments in Central Appalachia," in *Environmental Considerations in Energy Production*, edited by John R. Craynon (Englewood, CO: Society for Mining Metallurgy. and Exploration Inc. (SME), 2013), 214–27; Sara E. Sweeten, J. Sweeten, John R. Craynon, William M. Ford, and Stephen H. Schoenholtz, "Evaluation of Current Measures of Aquatic Biological Integrity in the Central Appalachian Coalfields: Efficacy and Implications," in *Environmental Considerations*, ed. Craynon et al, 381–94.

14. For example, Environmental Protection Agency regulations regarding water and surface mining require scientific expertise in order for industry to meet regulations. For one small sample, see: Environmental Protection Agency, Science Advisory Board, *EPA-SAB-11-2006: Review of Field-Based Aquatic Life Benchmark for Conductivity in Central Appalachian Streams*, https://archive.epa.gov/water/archive/web/pdf/epa-sab-11-006-unsigned.pdf

15. Shelia Jasanoff. *States of Knowledge: The Co-Production of Science and the Social Order. International Library of Sociology* (New York: Routledge, 2004).

16. Rebecca R. Scott, *Removing Mountains: Extracting Nature and Identity in the Appalachian Coalfields*, Quadrant (Minneapolis: University of Minnesota Press, 2010).

17. Deborah Stone, *Policy Paradox: The Art of Political Decision Making* (New York: Norton, 2001).

18. Alex Epstein is a favorite speaker with this crowd. Epstein wrote *The Moral Case for Fossil Fuels* (New York: Portfolio, 2014). He is criticized as a climate denier and of playing loose and free with facts (for example: Eric Gade, "Rebuilding the Ark: Alex Epstein and *The Moral Case for Fossil Fuels*," *BLOG//Los Angeles Review of Books* (August 4, 2017), accessed on July 13, 2023, https://blog.lareviewofbooks.org/reviews/rebuilding-ark-alex-epstein-moral-case-fossil-fuels/). Nevertheless, cheap energy has brought and enabled tremendous material prosperity and enabled the industrialization of Western society. Epstein is wrong, however, that fossil fuels must be the continued major means of cheap energy delivery. However, they remain the major players for now: (U.S. Energy Information Administration, "U.S. Energy Facts Explained." June 10 2022. https://www.eia.gov/energyexplained/?page=us_energy_home)

Natural gas—32 percent
Petroleum (crude oil and natural gas plant liquids)—36 percent
Coal—11 percent
Renewable energy—12 percent
Nuclear electric power—8 percent

Whether we find their use moral, we continue to use them for more than 75 percent of our energy output in the United States—not including what American-based corporations use for energy production abroad. Fossil fuel use continues to require scientific and technical expertise. Researchers in the fossil energy industry may rightly point to our gap in what we prefer to what we consume. See Emily Atkin, "Al Gore's Carbon Footprint Doesn't Matter," August 7, 2017. The New Republic, August 7, 2017, https://newrepublic.com/article/144199/al-gores-carbon-footprint-doesnt-matter versus what we actually consume.

19. Bruno Latour and Steve Woolgar, *Laboratory Life: The Construction of Scientific Facts, Second Edition* (Princeton, NJ: Princeton University Press, 1986); Robert E. Kohler, *Landscapes and Labscapes: Exploring the Lab-Field Border in Biology* (Chicago: University of Chicago Press, 2002).

20. Dorothy Nelkin. "The Political Impact of Technical Expertise." Social Studies of Science 5, no. 1 (1975): 35–54; Nelkin, *Controversy: Politics of Technical Decisions* (Beverly Hills: SAGE Publications, 1992). And, Nelkin, *Dangerous Diagnostics: The Social Power of Biological Information* (Chicago: University of Chicago Press, 1994).

21. Building on assumptions in STS and in particular, work in the Sociology of Science, that science extends far beyond the borders of a laboratory. Particularly illustrative to me in this vein was Robert E. Kohler's book *Landscapes and Labscapes: Exploring the Lab-field Border in Biology*. Not only does he examine a science which falls out of favor and finds favor again in terms of its status and resources devoted to it, but he traces the issues of producing science in the field rather than in a laboratory.

22. Geoffrey C. Bowker and Susan Leigh Star, *Sorting Things Out: Classification and Its Consequences* (Cambridge, MA: MIT Press, 1999).

23. See Endnote 2.

24. Robert K. Merton, *The Sociology of Science: Theoretical and Empirical Investigations* (Chicago: University of Chicago Press, 1972).

25. Shelia Jasanoff, *The Fifth Branch: Science Advisers as Policymakers* (Cambridge, MA: Harvard University Press, 2009).

26. *American Society of Mining and Reclamation*, http://www.asmr.us/#

27. https://www.osmre.gov/contacts/EmployeeDir/osmempdir.shtm?region=Appalachian#results

28. Judith Butler, *Bodies That Matter: On the Discursive Limits of "Sex"* (New York: Routledge, 1993).

29. Aaron M. Sakolski, *The Great American Land Bubble: The Amazing Story of Land-Grabbing, Speculations, and Booms from Colonial Days to Present Time* (New York: Harpers & Brothers, 1966); David Leadbetter, "Single-Industry Resource Communities: 'Shrinking,' and the New Crisis of Hinterland Economic Development," in *The Future of Shrinking Cities: Problems, Patterns and Strategies of Urban Transformation in a Global Context*, ed. Karina Pallagst, et al., 89–100 (Berkeley: Center for Global Metropolitan Studies, Institute of Urban and Regional Development and the Shrinking Cities International Research Network, 2009); Langdon Winner, *The Whale and the Reactor: A Search for Limits in an Age of High Technology* (Chicago: University of Chicago Press, 1986); Joseph Cowie and Joseph Heathcott, "The Meanings of Deindustrialization," in *Beyond the Ruins: The Meanings of Deindustrialization*, ed. by Joseph Cowie and Joseph Heathcott (Ithaca, NY: ILR Press,

2003); Steven High, "Capital and Community Reconsidered: The Politics and Meaning of Deindustrialization," *Labour/Le Travail* 55, 2005, 191–200.

30. Hugh Campbell, Michael Mayerfeld Bell, and Margaret Finney, eds. *Country Boys: Masculinity and Rural Life* (Penn State University Press, 2006).

31. David L. Brown, and Louis E. Swanson. *Challenges for Rural America in the Twenty-First Century* (Penn State University Press, 2010).

32. Donald Akenson, *God's Peoples: Covenant and Land in South Africa, Israel, and Ulster* (Ithaca, NY: Cornell University Press, 1992).

33. Indian Ministry of Rural Development "National Rural Employment Guarantee Act, 2005." n.d., http://rural.nic.in/documents/policies-acts-bills/department-rural-development.

Asia Briefing Ltd, Dezan Shira & Associates, n.d. http://www.china-briefing.com/news/2014/05/27/chinas-state-policy-rural-development-agriculture-industry.html; The World Bank, Europe and Central Asia Unit. "Monotowns: Path to Resilience: A Brief Review of International Experiences with Urban Regeneration," 2010.

34. Thomas Gieryn, "Boundary-Work and the Demarcation of Science from Non-Science: Strains and Interests in Professional Ideologies of Scientists," *American Sociological Review* 48, no. 6 (1983): 781–95; Giulio Lepschy, *A Survey of Structural Linguistics* (London: Faber and Faber Limited, 1970).

35. U.S. Census Bureau, "Metropolitan—Micropolitan," *Programs—Surveys*, July 13, 2023, accessed July 13, 2023, https://www.census.gov/programs-surveys/metro-micro.html; Leadbetter, David, "Single-Industry Resource Communities and the New Crisis of Hinterland Economic Development," *The Future of Shrinking Cities: Problems, Patterns and Strategies of Urban Transformation in a Global Context*, 2009, 89–100.

36. David Leadbetter, *Mining Town Crisis: Globalization, Labour and Resistance in Sudbury* (Black Point: Fernwood Publishing Co., Ltd., 2008).

37. Hinterland regions are places beyond micropolitan and metropolitan commutersheds. Hinterland residents are unlikely to commute to a micropolitan or metropolitan area to work for reasons of time, expense, or difficulty of travel. See Leadbetter, "Single-Industry Resource Communities"; Cristina Martinez-Fernandez and Chung-Tong Wu, "Shrinking Cities: A Global Overview and Concerns about Australian Mining Cities," in *The Future of Shrinking Cities: Problems, Patterns and Strategies of Urban Transformation in a Global Context*, ed. Karina Pallagst, et al., 29–36 (Berkeley: Center for Global Metropolitan Studies, Institute of Urban and Regional Development and the Shrinking Cities International Research Network, 2009).

38. Bruno Latour, "On Actor-Network Theory: A Few Clarifications," *Soziale Welt*, 47, 1996: 369–81.

39. Cynthia Duncan, *Worlds Apart: Why Poverty Persists in Rural America* (New Haven, CT: Yale University Press, 1999); William N. Parker, "Introduction: The Cotton Economy of the Antebellum South," *Agricultural History* 44, no. 1, 1970: 1–4.

40. See Emily Satterwhite, *Dear Appalachia: Reader, Identity, and Popular Fiction since 1878* (Lexington: University Press of Kentucky, 2011).

41. Ibid.

42. Erica Abrams Locklear, "A Matter of Taste: Reading Food and Class in Appalachian Literature," University of North Carolina at Asheville, Thirty-Fourth Annual Appalachian Studies Conference, 2011.

43. Michael Harrington, *The Other America: Poverty in the United States* (New York: Macmillan, 1962).

44. Shannon M. Woolley, Susan L. Meacham, Lauren C. Balmert, Evelyn O. Talbott, Jeanine M. Buchanich, "Comparison of Mortality Disparities in Central Appalachian Coal- and Non-coal-mining Counties," *Journal of Occupational and Environmental Health*, 57, no. 6 (June 2015): 687–94.

45. Rebecca Scott, *Removing Mountains*.

46. Benedict Anderson, *Imagined Communities: Reflections on the Origin and Spread of Nationalism* (Verso Books, 2006).

47. Reichert Powell, *Critical Regionalism*.

48. Jerry Bruce Thomas. *An Appalachian Reawakening: West Virginia and the Perils of the New Machine Age, 1945–1972. Vol. 12* (Morgantown: West Virginia University Press, 2010).

49. Chris Hedges, and Joe Sacco, *Days of Destruction, Days of Revolt* (New York: Nation Books, 2012); Herbert Reid and Betsy Taylor, "Appalachia as a Global Region: Toward a Critical Regionalism and Civic Professionalism," *Journal of Appalachian Studies* 8, no. 1, Spring 2002: 9–3; Herbert Reid and Betsy Taylor. *Recovering the Commons: Democracy, Place, and Global Justice* (Board of Trustees of the University of Illinois, 2010).

50. Travis Lowe. "I've Seen The Future. It Looks Like Appalachia: Life after the Job Apocalypse." *Medium*, March 31, 2017. https://medium.com/startup-grind/ive-been-to-the-future-it-looks-like-appalachia-5553bf89f275

51. For one example of this narrative, *see* Vannevar Bush, "Science the Endless Frontier." Office of Scientific Research and Development, July 1945. https://www.nsf.gov/od/lpa/nsf50/vbush1945.htm. As an aside, the imperative for knowledge production associated with science to also produce economic impact or benefit may have been an ill-conceived direction. Just as art must not necessarily produce economic benefit, science as knowledge production could also be justified as a cultural benefit. The recent direction has been to attempt to commodify the arts or humanities as technology's sidekick. I prefer to advance the concept that science ought to ally itself with humanities as also cultural "production" without having to purport definitive economic impact or assist corporate benefit.

52. For example, the "space race" between the United States and Soviet Union.

53. Thomas, *An Appalachian Reawakening*.

54. Ibid.

55. David Bloor, Jan Wolenski, and Matti Sintonen, "Sociology of Scientific Knowledge," in *Handbook of Epistemology*, ed. Ilkka Niiniluoto (Boston: Kluwer Academic Publishers, 2004), 919–62.

56. Wilma Dunaway, *The First American Frontier: Transition to Capitalism in Southern Appalachia, 1700–1860*, The Fred W. Morrison Series in Southern Studies (Chapel Hill: The University of North Carolina Press, 1996).

Chapter 1

Technology and the State of the State in the Pocahontas Coalfield

OVERVIEW

How did we get to this state of Big Rural, of the use of rural areas in the United States locally and globally unacknowledged in broader culture and politics as sites of industrial production yet coproducing a system "too big to fail"? The history of the process of industrialization of society and the history we are taught of it, if taught at all, in the United States focuses on factories, Fordism, Taylorism, production in mills of the U.S. Northeast, and the streamlining of production of cars. Cities had factories and city slickers, and rural areas had country bumpkins, sitting under trees napping, smoking a corncob pipe, or chewing hay straw. The leaving of, or, reshaping of, agrarian life becomes a footnote. Industriousness and industry happen in urban areas, or at least in the narrative of progress.

The growing of, or the extraction of, raw goods to feed industry becomes secondary or obscured in comparison to the factories converting them for further industrial use or consumption.

This tale of urban industrialism also subverts the fact that industry also was developed in many rural places, that this extraction of raw goods from the rural for urban factory use became increasingly streamlined through science and engineering located in and occurring in the rural, and that this intense development also greatly separated nearly colocated rural places and the people inhabiting them. As a matter of personal narrative and illustration, by the 1940s my father as a child was taking the bus from near the farm his miner father "retired" into owning, and getting out in Bluefield, in Mercer County, West Virginia, a town built for the middle and local managerial class developed in response to the industrial work on the coal seam in the Pocahontas Coalfield (straddling the West Virginia–Virginia border in neighboring McDowell

County, West Virginia and Tazewell County, Virginia). By this time, Bluefield had a twelve track railyard, Art Deco buildings, a youth Drum and Bugle corps (which my father was ostensibly attending but instead was spending the day in movie matinees). During this same period, my mother one county away in Bland County, in the Virginia mountains, toiled with her family as a sharecropper, farming by hand or by mule, in a rural place where electricity was not yet available. By the time she was a young adult, a Sealy mattress factory and a couple of sewing factories had opened, one near where her family had farmed, and another over the mountain in West Virginia, and she went to work in those. And while, to reference Harry Caudill, *Night Comes to the Cumberlands: A Biography of a Depressed Area*, that some rural parts of Appalachia perhaps never recovered from their pre–Civil War prosperity dependent more on a kind of frontier mode of living and working, other parts, with the development of industrial-scale resource extraction, experienced cycles of boom and bust, their local economies tied to national and international interests and events as much as those of any urban industrial space.

Through the promotion of the Pocahontas Coalfield rural space for industrial development and capital investment, its science and technology narrative, by proxy, may also foretell our modern relationship in the United States with what we conceive of as rural. The Pocahontas Coalfield's current social and environmental state, with high poverty and many brownfields, results from a mix of corporate and academic science and technology research toward economic efficiency and the pitfalls of the single-sector rural industrial space as a simultaneous work and civic space.

The Pocahontas Coalfield is NOT a separate world, an Other or (though maybe Othered) place, somewhere out there, not connected to the larger picture or economy, disconnected from globalism or international markets, isolated from international and sophisticated science. The Pocahontas Coalfield, in its most recent economic stage, encompasses a place of complex poverty *and* complex science—a single-sector rural economy tied to multifaceted international companies. It also remains a place constructed *on purpose*, whose original raison d'etre has veered in the opposite direction from that envisioned by its first and maybe fiercest conceptual advocates. In this latter respect, in addition to all the other cautionary tales the Pocahontas Coalfield embodies regarding rural industrial single-sector economies, myths of self, place, and work, predictions of machines and men, it also reveals a tale of being careful of what one wishes for—or, of the folly of good intentions and the foolishness of predicting the future.

Just as varying kinds of masculine performativity exist, so do many ways to perform the rural. They overlap with geography in attempts to quantify and qualify whether a space can be construed as rural. Some definitions rely upon a count of inhabitants; others rely upon the presence of certain kinds of ways to make a living such as agriculture and/or mining; others quantify or qualify the presence of nature, whatever that is, as a kind of rural space in contrast to the metropolitan.

What the field of Science and Technology Studies of Society (STS) brings to this qualifying or quantifying of the rural may be a 20,000-foot view of the process of classification and how classification becomes a means of control or for control. Rather than only a scientific classification of tools or flora and fauna, the social constructivist position of and the awareness in STS of classification as a human-driven activity highlights for us how we take for granted that classifying and categorizing are usual, right, what ought to be done. Whereas cultural studies, rural studies, and so on may be concerned with more accurate classifications of the rural, or as rural studies may now typify the rural as *rurals* (meaning that no one single kind of rural exists; a range of rurals happen simultaneously), an STS perspective asks to begin one step prior in having us consider what compels us to classify, that is, what do we hope to gain through more "accurate" or "better" classification of the rural?

One effect of following patterns and making links could be to impose a hegemonic narrative of the rural, squeezing out the rurals that do not fit into the linkages I make. The goal of this accounting is not to produce a monolith. There are many rurals, a range of factors contributing to rurality. Big Rural can also shift, flex, pulse, contract, expand, include, and exclude, as can other kinds of "too big to fail" modern entities such as "finance," "banks" or big corporate catch-alls such as "Big Ag" and "Big Science." I point to connections as I make sense of them and as the ground situation for Big Rural stands within my view and within my present. The connections I shed light on may illuminate cultural and technological territories where other kinds of gender, racial, class, ethnic, and hegemonies already tower or assert. Because one set of descriptors or connections cannot be all inclusive because one person is incapable of seeing every set of possibilities all at once neither indicates an active desire to exclude nor does it mean that linkages should not be attempted. Thus, if in my links I exclude, and you the reader can readily point out my exclusions, then you will be producing a map of where I have been blindsided by hegemonies or dominant narratives. I welcome the interest in pointing out the potential for connection that I have not otherwise indicated, and I welcome the chance, too, to engage in thoughtful consideration regarding the factors I came to focus on for to depict Big Rural as a technological system containing also local and localized variation.

I problematize the rural coalfield space, revealing technology and social-related (socio-technical) issues. I open the black box (a concept not examined, but just used without critique; one such term is the "economy"—often used, but not defined) on some kinds of persistent Appalachian poverty and on coalfield automation, resource depletion, and the United States' development of big Ag (large scale agriculture achieved through larger machines and industrial scale land manipulation with massive chemical and fossil fuel inputs) in the Midwest and West over the smaller acreage family farm prevalent in Appalachia until the 1950s. I examine the contributing factors to a democracy deficit (aka low voter turnout, low political participation, low civic participation, a life controlled not by citizens but by corporate or government forces) in thc coalfield space, and single-sector economic spaces more generally. I address the single-sector rural industrial space and the potential issues of corruption and depleted social capital (not only do we have money capital in our lives, but we also have the "credits" and "connections" we can spend or expend with other people, or, as a community,to advocate for what our community needs). I argue against the concept of "resource curse" (that places where the economy is based on one single fossil fuel resource somehow cannot also be democratic), terming it essentialist (a fact of nature), preferring instead to address the problem of corruption in the fossil fuel rural industrial space as correlated to that in single-sector economic spaces more generally, whether rural or urban. I discuss the trends in the qualitative interviews I conducted in the Pocahontas Coalfield and adjacent rural spaces with a focus on "lack of leadership" as the top problem identified by regional respondents before issues of work or health and define this problem as a function of the single-sector democracy deficit.

Poverty's Persistence Has Roots and Reasons

In *An Appalachian Reawakening*: *West Virginia and the Perils of the New Machine Age*, 1945–1972, Jerry Bruce Thomas succinctly summarizes the state of the state and the contradictions at work in much of West Virginia beginning in the 1950s. As the rest of the country enjoyed never-before-experienced levels of prosperity, shifts in United States' energy use, along with changes in coal mining technology, created more poverty than prosperity for coalfield workers. Thomas characterizes the paradox for the single-sector coal sections of central Appalachia, and the single-sector agricultural sections of Appalachia in this way. For single-sector coal sections, shifting energy demands toward natural gas and oil coupled with technological change affect in the coal sections of Appalachia the same "dilemmas and dangers for workers and residents" yet, in other sections of the United States, though workers provided "sweat, muscles, and backs" for economic growth, they also enjoyed some of the fruits of this new prosperity.[1]

Though Thomas does not make this characterization, I contend that in some other sections of the United States, Joseph Schumpeter's "creative destruction" functioned—new technologies replaced the old, thereby also creating new economic and work opportunities in the same or approximate location.[2] In Appalachia, gas or diesel uptick in industrial use and machines built to supplant miners reduced the need for workers mining coal. For the most part, little new technology as work arose like a Phoenix in the same locale for those thrown out of work due to these shifts. Likewise, according to Thomas, to keep the coal industry afloat, the United Mine Workers of America chose to side with coal management with respect to the health risks associated with these new machines.

As an illustration of the trajectory of growth stagnation and economic stagnation, overall, in the 1950s, the United States added 15.5 percent more workers, and West Virginia's rate of employment fell nearly as much (14.3 percent). Thus, West Virginia had a nearly 30 percent lag in employment behind the rest of the United States, on average.[3]

Important to note, Thomas highlights a general U.S. fear that has largely disappeared from robust national discussion: automation. In 1960, more Americans feared "automation" than feared the Soviet Union. Automation was understood to undermine high-paying work.[4] While deep mine safety had been a major impetus for the creation of the United States Bureau of Mines, new machines in use in deep mines beginning in 1948 saw the rate of mine injury increase—almost doubling.[5] With these new machines, workers in the mine had to pick up the pace, opening more possibility for injury, and creating a situation in which miners could not hear changes in stress to the roof. The amount of dust in the air also increased, which then led to more lung complications, the full extent of which is only coming out in recent documentation regarding true numbers of miners sick from silicosis, coal worker's pneumoconiosis (Black Lung), and cancers.[6,7] More dust also meant even less visibility. In an already gaseous environment, the dust could make the machines volatile.

Strip mining slowly caught on in Southern West Virginia during World War II, as the federal Solid Fuels Administration pushed for more coal production. Until then, according to Thomas, environmental and aesthetic concerns had overridden interest. Post–World War II, strip mining proved more economical, both in production and in reducing associated labor costs. Large machines such as power shovels, draglines, bulldozers, and equipment associated with road building replaced underground mine workers. In addition, local towns and municipalities had to contend with blasting at the sites, rain that moved acid mine drainage into streams and wells and that ruined plants and trees. A long battle for regulation and reclamation of strip-mining sites, begun in the 1930s and culminating in a law in 1967, ended with the West Virginia Department of Resources swamped in their attempts at enforcement as companies increased surface mining production.[8] In the 1970s, efforts in Kentucky and West Virginia to halt strip mining ramped up. By then, the clear disadvantages for mine workers prompted one West Virginian to note to the *Charleston Gazette* that most of the strip mine workers were not locals and their skills could afford them opportunities in other industries.[9] A long-standing democracy deficit, however, led to the legislative defeat for a ban, as many political representatives either actively worked or owned in coal or had significant supporters in coal.[10] Furthermore, the agency tasked with enforcing the 1967 law could not keep pace in financial support nor backroom deals with coal industry advocates, including sitting governors.[11] A ban on strip mining was defeated, and, regulations banning or reducing the impact of more aggressive means of strip mining were rolled back.[12]

Agrarian Technological Shifts and Remaining Timber and Extraction Issues

Jerry Bruce Thomas addresses a second machine age derailment of work, this one in Appalachia's agricultural sector during the 1950s.[13] Just as coal mining employment entered a free fall post World War II, mountain agriculture could not compete with the large machines and surplus inputs or agricultural political advocacy of the American Midwest and West.[14] Machines designed for flat land and opened (not open, but opened deliberately to Western agriculture) spaces did not suit mountain agriculture. As commercial agriculture shifted even harder to large tracts of lands, smaller holdings of fifty-five acres or less in Appalachia could not compete neither American tastes nor agricultural policy focused on supporting agricultural product diversification or small farm advantages.[15] This compounded the shift prior to World War II of farm offspring leaving the farm, prior full-time farmers finding employment off the farm, and farmers aging. Additionally, erosion, mostly from timbering, had wrought severe soil deficiencies in some 25 percent of available farmland in the West Virginia mountains.

Thomas points to some throughlines of poverty in the region not only being caused by coal industry mechanization and shifts away from coal energy, but also to the view at the time that small farms could not "be enlarged to commercial size."[16] He also relates other natural resource extractive or environmentally destructive industries as equally devastating for some West Virginia farm communities. For example, in Hardy County, tanneries and timbering had left that area susceptible to severe flooding. As these industries closed or reduced workers, off-farm work dwindled, and landowners could no longer afford to even farm part-time. As in coal country, as mechanization happened (in timber, or as industry demands fell domestically as in tanning and textiles), those who did not move for work elsewhere, or could not work elsewhere due to age, disability, family demands, and so forth found their standard of living drop to below benchmarks for poverty. Improved approaches to forest management addressing issues of erosion and forest planning failed to replenish former timber jobs or lumber demand.

Thomas points out the obvious, that the cities and towns tied to these industries and the many businesses formerly serving the employees or businesses themselves also declined with the drop in demand for workers. Population decline set in, and this population decline rippled and roiled through many once prosperous towns and counties, contributing on multiple levels to economic decline, education decline, and tax base drain.[17]

Boom and Bust of Coal Demand and Employment

In his 1992 work, "Addictive Economies: Extractive Industries and Vulnerable Localities in a Changing World Economy," rural sociologist William Freudenburg analyzes the dearth of entrepreneurship in the United States Appalachian Coalfields in particular as being related to the cycle of boom and bust of mining: even during hard times for the industry, there was always the promise of a job on the horizon—someday the mines would start back up, so you just needed to sit tight till then. Sit where you are and wait; good times were on their way.

Freudenburg likened this riding out of boom and bust cycles to the cycle of gambling addiction, what he asserted to be a very powerful lure for the next "hit" of work time, industry upswing, and prosperity.[18]

Despite this lure, coal counties across this region have declined steadily in population since the advent of the continuous miner in 1954.[19] Even at that time, the outlook for employment in coal was bleak as a special report for Virginia predicted and traced the United States' move from coal to oil dependence. Prior to the continuous miner, coal already was losing market share and shedding jobs.[20]

Democracy Deficit, Corruption, and Decline of Social Capital

Participation in the market economy can be hindered by geography. Areas beyond suburban or exurban geography may not be capable of keeping pace with more central locations in terms of economic growth. In the United States, participation in the market economy and economic growth have become associated with promoting economic stability—all feature goals of U.S. policy at home and its policies abroad. On the one hand, resource extraction is used by industry to tout job creation in hinterland, or less geographically central, areas.[21] On the other hand, academics and researchers propose that resource extractive industries squeeze out the possibilities for the growth of other kinds of industry and stifle entrepreneurship and economic diversification.[22]

One of the key concepts in a liberal democracy rests with the idea, and the ideal, of being a free agent, not bound to a particular person or bound to the land. One is free to choose where to live, for whom to vote, and so forth. In a town that depends mainly or solely on one industry for its livelihood (a *monotown*), the issue of remaining a free agent and being able to practice democratic citizenship can become difficult.[23] In the U.S. context, if where we work is not unionized or nonhierarchical, we anticipate that democracy is what we practice when we leave the workplace.[24] At work, owners run the workplace top-down, with less or more adherence to labor laws depending on the local level of enforcement.[25] One contention claims that competition for labor in urban geographies reduces the impact of business on the private lives of citizens.[26] Ideally, in an urban context in a democracy, one can vote, attend a protest, gather, and so on, and otherwise exercise democratic practices with respect to political governance without direct interference by one's workplace, because the workplace does not monopolize local government.[27] By contrast, in a monotown, where one's workplace directly influences one's place of residence as the town exists primarily to service this one industry, the totalitarian rules of work may obtain and control of local politics may be held tight directly, or indirectly by the mono-industry.

Resource extraction was once heralded as a certain road to material prosperity.[28] Not only did it support boom monotowns, but in the United States, it also set the stage for physical work stoppages by labor, thereby making workers more efficacious in their demands for fairer wages and more humane working hours and conditions.[29] In particular, in the early twentieth century, for example, coal mining's structures and heavy dependence on a large human workforce made it susceptible to labor's power to bring production or distribution to a halt. In the United States, as the twentieth century progressed, mine operators shifted from heavy labor to heavy technical reliance, thereby eliminating and/or reducing the threat of labor stoppage.[30] Many resource extraction locations continued to be highly productive, yet with a greatly reduced workforce. The human aspect of monotowns that had boomed with initial extraction shrank without any plan for how to accommodate the problems that attend shrinkage: loss of necessary businesses, loss of taxes, emptying schools, crumbling infrastructure, youth and brain drain, and so forth.[31]

Unlike the promise of renewal under economist and political scientist Joseph Schumpeter's theory of creative destruction (which, in a nutshell, maintains that in order for capitalism to thrive, old processes get destroyed by innovations, thereby reinvigorating capitalism as a system), these extractive communities, as *communities*, often shrivel with no other active or activated economy to offset that of extraction.

Even an entity as contentious as the World Bank asserts ideals about what makes a community a community: the basic obligations of a local government to "ensure the delivery of quality social and communal services and a responsive supply of land and housing, so that the city becomes and remains a hospitable place for businesses and households."[32] Local government can and should function as an essential counterforce to significant private companies. By contrast, extractive companies often are "footloose, making little lasting investment for the town's general benefit, and sharing profits only minimally through local tax revenues."[33] For extractive communities, by the time an extractive industry reduces the human workforce either through increased mechanization, reduced need for the target extraction, or depletion of the target extraction, little economic counterweight exists (if it ever did).

According to historian Timothy Mitchell in his work *Carbon Democracy*, the kinds of communities that historically became sites for extending democratic practice through massive unionization and strikes often, later, may become less democratic than places in society's centers influenced by their democratic practices (however, there are theories that these sites never realized true democratic potential due to the edging out of other industry by the mono-industry). For example, by the specific elimination of a human workforce through the introduction of technologies, management reduces the potential for work stoppage—the backbone of what lends a strike its power.

If, in these cases, local government consists of members loyal to the company or local industry, and, industry also reduces or has reduced the human workforce through mechanization, a monotown ceases to deserve the moniker "town." In the West, in the context of modern democracies, to constitute being a "town" requires that some political power remains within the grasp of local citizens through local voting, community boards, municipal elections and positions, etc. If both political force and economic power in a town belong to industry, then, for all practical purposes, the town's function has been reduced to that of an extension of the company. The town becomes (or remains) a kind of department responsible for, among other things, providing entertainment, food, shelter, and education for the place-dependent human technology needed by industry.

With respect to the monotown's role in a larger democracy, the peripheral monotown continues to provide many of the resources needed for society's centers (metropoles) to exist. Without cheap energy, most of the goods and services affording residents of more metro centers the life chance to pursue nonresource extractive, nonagricultural, or nonindustrial professions would not exist. The move over the twentieth century to abstract, education-based (*elite*) conceptualizations of democracy away from democracy as practiced by blue collar workers at that century's start further undermines the power of democracy as a practice rather than as an ideal.[34]

As for the democratic struggles of workers or laborers not exposed to higher education? Democracy relies on democratic practices such as collective bargaining, fighting for workers' rights, mobilizing for voter enfranchisement, or participating in concrete acts toward social change. These kinds of practices of democracy outweigh in value abstract analyses of democracy.

On these points, the practice of democracy rarely extends from the polity into the company unless through the collective action of the employed. Therefore, if a monotown has become a technological extension of the industry, that it is not democratic may neither be obvious nor problematic to its inhabitants. If the "town" functions as or extends the industry's technology and workplace mores, the town's inhabitants may first construe themselves as necessary first to maintaining the flow of company products and then, if at all, also as citizens deserving of a voice and deserving of impacting their lived environment.

It is important not to succumb to discussing companies as if they, too, remain static and do not consist of people. In larger discussions in the United States, we often discuss corporations as if they were some kind of nonmaterial entity and as if decision-making processes did not ultimately rest in the hands of actual people. Though legally corporations can stand as entities separate from individual shareholders or employees, ethically they should not.[35] Ethically, the people of which the corporation consists become citizens granted legal permission to suppress the rights and democratic livelihoods of other citizens. Masking the human agency in this process may be pseudoscientific economic data representing profit-driven human agency as mathematically derived technological efficiency.

As profits and economic impact can be calculated, their formulation and representation assume the "virtues of necessity and universality generally attributed to scientific rationality."[36] To be concerned also with human actors in the technological space of the technical monotown pejoratively demonstrates overly emotional or irrational concerns. Furthermore, if the technical monotown produces commodity goods such as coal, timber, natural gas, oil, grain, etc. to suggest alternative concerns other than economic efficiency can be construed by industry and by government as an aim to cripple national interest, thereby to risk being unpatriotic.[37] Economics or quantitative economics depends upon actual human agents. Economics functions as a disciplining system on competing goals for the technical monotown. The possibilities for choices not adhering to economic or pseudo-technical proposals become curbed.

This last term—"economic" along with "the economy" or "economy"—are terms that are often black-boxed; politicians, economists, journalists, policy-makers, and so on utilize these terms without defining what they mean, taking for granted that these terms have fixed definitions, when in fact the terms and the range of concepts they represent exhibit a range of socially created and negotiated practices.[38]

Usually, in contemporary use, these cases are directly couched in the language of cost benefit: computer programs are more accurate than the human white-collar workers they replace and less costly to maintain, and so on it goes with Schumpeter's creative destruction.

To sum up, hinterland geography can impair a diverse capital-based economy. A diverse economy can contribute to diverse politics.[39] If you cannot speak against the political power in your community for fear of losing your job (if the company also runs the politics), then you do not have freedom of speech. Not as many people become needed in many monotowns due to technological shifts such as automation—thus, not enough people exist to counterbalance the industry's power. Under neoliberalism, freedom has been subsumed under the free market to where the latter now embodies also the former. This leaves democracy an abstract concept monopolized by elites rather than as a set of practices accessible to all citizens.[40] We also speak of corporations in the abstract rather than the sum of actual people making decisions, thus people in corporations are not held as directly accountable as are individual citizens. To speak or act against a commodity of national interest can make citizens in the monotown appear unpatriotic.

The Limits of Utility for the Concept of the Resource Curse

The materiality of technical artifacts impacts the social; in fact, we may even posit that these technical artifacts coproduce the social. "Steel, wires, semiconductors, nuts, and bolts" may impact how people live, what choices they can make, and how they contend with each other in the polity.[41] This assumption underlies the concept of the "resource curse."[42] According to the British economist Richard M. Auty, the presence of natural resources in a newly developing area creates a vacuum of power and stifles economic competition such that ruling elites rely upon rents from natural resource extraction to maintain political power rather than submitting to the will of competing economic interests or to the will of the people. The resource prescribes local economic possibilities. Likewise, a similar phenomenon, "Dutch Disease," occurs even when a newly extractive country implements careful policies, but still can become affected by inflated exchange rates and wages due to an influx of extractive rents. In this case, the factual materiality of a natural resource predicts the political outcome.[43]

Carbon resources do not guarantee prosperity at the sites of extraction. No material technology guarantees material prosperity just as the fact of having laws or policies (technologies of government) does not guarantee they will be followed. In fact, we have come to anticipate some gap between laws or policies and their adherence.[44] Certainly, fear of legal action or other state-directed retribution for nonadherence can make people toe the policy line; however, other modes of governance and civil society advocacy also lend policy weight. That one can get to a particular building wherein particular people work on a given issue may serve as a point of relief or stress (depending on the social standing) regarding a particular issue: think the political and social power of a union hall. The embodiment of accepted practices through the creation and maintenance of institutions empowers these institutions as disciplining or "policing" forces.[45] For example, the creation of a regulatory agency may lend material weight to the directives to follow a particular set of policies—but, as Jerry Bruce Thomas points out in the case of strip mining enforcement in the State of West Virginia in the 1970s, policy ranks only as effective as it receives institutional or other resources for adequate enforcement.

Prior to an institutionalized labor union for miners, for example, the union for miners lived not in a particular building or site but first in particular practices, and only when people enacted those practices. Yet, once the union for miners assumed institutional trappings such as a building, and professionalized, a miner no longer needed to practice the union to experience the union. That miner could go to the material building and seek an expert to address grievances which heretofore may have only been expressible through forms of direct democratic action.[46]

So, then, material carbon resources also receive their punch from the same source that gives laws or policies theirs—configurations of the social. In "Correlates of Corruption: Rethinking Social Capital's Relationship with Government in the United States," Richard Ledet paraphrases other scholars and agrees that "scholars interested in U.S. Politics have" "remained steadfastly uninterested in corruption issues for generations." To put this into relief, though the field of economics has long focused on cost-benefit analysis and other idealized means for predicting the efficiency and efficacy of potential policies, the same amount of energy for the rational ideal has not been applied to uncovering empirical relations or for producing an acceptable state of corruption.

So, then, what of corruption in one of these U.S. extractive states? In highly extractive West Virginia, with significant rents from coal, gas, and oil, why do many people in the most coal-producing counties suffer the largest poverty, consistent with the myth of the resource curse? Jerry Bruce Thomas demonstrates that this poverty was not only due to fossil fuel extraction, but to other kinds of extraction such as timber, and that other kinds of environmental degradation, such as tanneries, also took their toll on employment, and the mechanization of these industries led to massive employment and economic loss. In these cases, being an area of mono industry affects corruption more than the fact of the area's being extractive. Upon closer examination, though West Virginia's most extractive counties rank in the middle of the state's fifty-five counties in terms of per capita income, over all the state's lowest per capita income *towns* are in its most extractive counties. Furthermore, while West Virginia may rank fair to poor along with many other U.S. states in terms of political corruption, it ranks second from the bottom in poverty.

How does it then perform corruption?

Ledet differentiates among variables within the construct of social capital. For example, *networks* (his emphasis) "provide the human infrastructure necessary to accomplish economic, political, and social goals." Furthermore, *values* (again his emphasis), "or norms" can be constructs such as interpersonal trust and political equality. In a study he cites in Iowa, communities with "high levels of social capital also have high-quality governments." Internationally, "higher levels of social capital are associated with lower levels of corruption." Ledet further points out that from the research on good governance, he concurs that social capital as defined as trust strengthens it as a construct and highlights the role of trust in good governance.[47]

Democracy becomes precarious in the single sector town due to the conflation of work and civic life; adding corruption to the mix further undermines democracy in that it "breaks the link between collective decision-making and people's power to influence collective decisions through speaking and voting, the very link that defines democracy."[48] Perceived corruption can cause people to lose faith in government as a technical entity, regardless of whether the actual technical government entity is still or actually corrupt.

Ledet points out a muddying factor with respect to extractive communities. If the number of political prosecutions defines corruption, then whether someone who is politically corrupt in fact comes to trial and is prosecuted varies widely from state to state. This, then, renders a comparison of corruption based on prosecution as minimally useful. More robust than conviction are public perceptions of "the abuse of public office for private gain." For example, the use of journalists as reporters of corruption concurs with methods in the corruption literature. Researchers view journalists as well-informed and well-positioned to judge perceptions of local or state corruption.

Case in point, in *Afflicting the Comfortable: Journalism and Politics in West Virginia*, Thomas Stafford, a former investigative reporter, chronicles the corruption of West Virginia's politics from 1945 to 1993. Though much of the text traces the corruption with respect to governors and vendors, Stafford also follows the boom and bust of the influx of money into the region through anti-poverty programs and how these programs lacked oversight (more on those below in the section on Mingo County and Huey Perry). The struggle between business interests and miners marked the government as being in the pocket of business until West Virginia Senator Robert Byrd's election in 1953, with the first prominent former impoverished person of coalfield heritage entering the larger arena of state-level politics.

Stafford also contends that resource extraction during World War II took an additional toll on West Virginia, through strip-mining, and links these corporate interests to the same corporations responsible for fiscally reckless practices that had led the country prior to that into the Great Depression.

Furthermore, Stafford outlines several conundrums of West Virginia and politics including West Virginians' loyalty to a politician, even if that politician is revealed to have been highly unethical and highly influential in his/her corruption. He cites the case of Governor Arch Moore, indicted in the early 1990s for bribery, tax evasion, lying under oath, fraud, and buying votes. Despite his long history of corruption and leading by poor example, many West Virginians still viewed him favorably and commended his rule (and elected his daughter U.S. senator, which she remains as of this volume's publication, and, other members of the Moore family also anticipate becoming or already voted in). By contrast, Stafford quotes U.S. Attorney Michael Carey on Moore, "Those that knew him [Moore] said he had been a crook for years, and public officials seemed to believe that everyone was doing it [corruption], including the governor, so why not them."

Between 1984 and 1990 Carey prosecuted sixty-nine county, state, and federal officials. Though it could be contended that with this scandal, corruption in West Virginia came to its peak, Stafford reasons something less joyful. He notes that many of West Virginia's public servants had become cavalier in their acceptance of corruption. In 1991, "members of the legislature claimed that the ethics law they passed two years earlier was too rigid. They were being denied . . . the opportunity to accept hotel bonus points . . . choice tickets to university football games . . . white-water rafting trips . . . [which] put them at a disadvantage in administering the affairs of office." He also notes the firing of a state employee who filed an ethics complaint against management and the investigation into public servants' utilizing state resources for themselves.

Stafford further points out that the kind of investigative journalism that may have revealed Moore's corruption earlier can be costly and that only a handful of West Virginia papers have the staff and legal support to conduct lengthy or potentially litigious reportage. In recent years, the will to conduct these investigations may have entirely disappeared with the buying up of the local remaining presses by individuals with fossil fuel interests.[49]

On this same strain, Huey Perry's *They'll Cut off Your Project: A Mingo County Chronicle* reads like an extension of one of Stafford's chapters, though Perry holds a contrasting opinion of Robert Byrd. Stafford counted him as a friend and viewed him as an ally of working people. Perry viewed Byrd as an accomplice to cronyism, and indeed, someone that chose not to intervene on behalf of working people in the following incident. In *They'll Cut off Your Project*, Perry chronicles how he (Perry) came to head anti-poverty programs in Mingo County,[50] a major coal producing county, in the mid-1960s, and how his five years there played out. In addition to deeply involving the poorest people in Mingo County in the management and decisions of the Office of Economic Opportunities (OEO), Perry et al. took on local politicians and politics with respect to rampant election fraud. Perry posits that the smear campaign local politicians and business-owners ran against the management of OEO had to do with these politicians and businesspeople not being able to dip into the anti-poverty money used for OEO. Moreover, with the direct involvement of poor people in OEO decision-making and action-planning, Perry et al. countered the prevailing concept that mountain people would not stand up for themselves. However, by 1969, after challenges for potential takeover of OEO by the Mingo County Court, the OEO began to receive directions from its federal sponsors not to involve the poor in decision-making, and, to include "all segments of the county in policy decisions and the administration of community action programs," in other words, politicians and businesspeople. If the people involved with OEO demonstrated or protested against elected officials, OEO's federal funding would be withdrawn.[51]

In Perry's case, the OEO came up against the wrath of a local political boss, State Senator Noah Floyd, who also went to bat for Governor Arch Moore during the anti-strip mine abolitionist rallying of Moore's campaign for a second term.[52] Throughout his book, Perry demonstrates the intense local political plays and interests in anti-poverty money as well as examples of strong-arming, cronyism, and election fraud by local politicians. A final demoralizing blow came when four local politicians accused of rampant election fraud and vote-buying were acquitted, despite abundant testimony, leading a U.S. Attorney from the Justice Department to remark regarding Mingo County, "Freedom has been lost. . . . There is a government of the organization, by the organization, and for the organization. . . . The only thing people get when they go to the polls in Mingo County is money."

Adding Stafford's chronicling of gubernatorial corruption to Perry's chronicling of that at the county-level is Cynthia M. Duncan's characterization of patronage politics in her Appalachian amalgam site "Blackwell." In this coalfield area in Appalachia, the school board functions to provide jobs rather than education. Indicative of how this and other corruption may affect the coalfields are the words of a local educator in Blackwell regarding politics, "You can't break the whole web. You can maybe break strands of it, and you hope, somehow, when they get rewoven, that they're different people with different ideas and things change a little."

Duncan points out that this corruption accompanied coal operators to Blackwell. As a means of contrast, some theorists of the resource curse point out that Norway, long before its oil boom, already had developed into an advanced democracy (though, this proposition could be questioned), thus, it entered the extractive resource market well-armored in social capital to contend with corruption.[53] However, if taken at face value, the rule to extraction would be that democracy may only be possible in resource extractive areas that were already stalwartly democratic prior to the development of an extractive industry. An important side note regarding Norway: while itself democratic, it has had millions of acreage in natural gas interests in Appalachian West Virginia, Ohio, and Pennsylvania, where it did not contribute to strengthening local social capital or governance.[54]

In terms of creating the situation and culture for coalfield environmental, social, democratic, economic, and financial degradation, all contributing factors cannot be accounted for, yet corruption and technological advances in the coal industry intertwine and coproduce. They not only historically collide; they continue to resonate with current change agents and social service providers working on or in the Pocahontas Coalfield and other adjacent coalfields.

Problems Now, Yet Lack of Leaders Ranks First

The environmental, social, democratic, economic, and financial degradation of the Pocahontas Coalfield can be well-documented in current statistics from the Centers for Disease Control, the U.S. Census, state assembled statistics, various academic journals, and through investigative journalism. In addition to the major issues outlined above, a few additional factors make the creative destruction and economic rebirth of this region different in profile from cities going through restructuring after the collapse of the need for workers in their single-sector economies (think Detroit, MI, or Buffalo, NY)[55]:

Unlike mass exodus or abandonment of land and buildings in major cities, the extractive companies are not letting loose many of their coalfield holdings or mineral rights. Someday coal (or something derived from it) may return, in the meantime, however, rare earth minerals and natural gas may be developed. Thus, entrepreneurs looking to utilize the land in this region must focus on what they can access by way of lease agreements on the surface. Furthermore, accessible land also must be ascertained for viability considering longstanding chemical contamination from the energy or railroad sectors or industry-incited erosion.[56]

The heart of this region sits well beyond commutersheds, that is, most of the southern West Virginian and southwest Virginian coalfields sit far from large cities, thus, little opportunity exists for converting a town there into a bedroom community.[57]

Many people in the Appalachian coalfield and the coal adjacent counties are unaccustomed to traveling outside their area for work and fear navigating even smaller local micropolitans. One qualitative interview respondent discussed the issues of finding jobs for locals in which 30 miles away was not local enough, and, his clients feared travel.[58]

Local education remains focused on preparing young people for jobs that do not exist there (in the energy extractive industry) or preparing them for export to other regions of the state or country.[59]

The remaining population has become disproportionately:

a. Elderly[60]
b. Disabled[61]
c. Opioid addicted[62,63]
d. With households headed not even by grandparents, but by great-grandparents as parents and grandparents contend with addiction.[64]

As discussed previously low social trust[65] and higher poverty[66] obtain.

Furthermore, race and racism have never been dealt with adequately in the coalfields. There is a long and complicated history of Black middle-class success, marginalization of Black culture, and active racism.[67]

With the region's turn toward Republican voting, it is now often politically at odds with programs that provide relief.[68]

Access to services remains difficult, with elderly and disabled people not able to descend from their mountain homes to get to services, even if they live in town on a ridge.

A local workforce can be difficult to hire and maintain due to addiction and up to 60 percent of children in McDowell County do not live with their immediate family, but with grandparents and great-grandparents, and, are at risk of becoming wards of the state when those relatives pass. These elderly relatives have issues with having limited incomes and being disabled.

Land ownership largely remains in the hands of corporations.[69]

Land also can wildly vary in its viability for economic development beyond energy due to brownfield or other contamination.[70]

Words and Thoughts of Grasstops in the Pocahontas Coalfield

Given the more historical nature of most of the scholarship in review for *Big Rural* and given that few scholarly sources spring directly from coalfield or coalfield adjacent residents themselves, for this work I chose also to pursue qualitative interviews with grasstops respondents working on a range of issues and projects in coalfield and coalfield adjacent counties. I wanted to work from their knowledge of the on-the-ground situation and, I wanted, as a locally engaged researcher, to hear directly from people working and living every day in the geographic region of concern. Additionally, as a participant observer, I attended community improvement meetings, regional and national conferences addressing economic issues in Appalachia, and spoke with a range of grasstops working more generally on economic, community development, and environmental restoration in Appalachia.[71]

Four trends arose from these different grasstops sources. Attending conferences focused on improving the region, one main issue of concern arising was creating economic opportunity or entrepreneurship. Prior to my interviews in the region and my participant observation, I had anticipated that this would be the major theme. However, in the region of concern itself, as I interviewed grasstops, a different major theme appeared: lack of leadership/lack of vision—which my respondents often intertwined.

I had fully anticipated that the effects of poverty or the effects of coal on the environment would rank as the top concerns, given the scholarship I had read. By contrast, two respondents listed for me the opioid crisis nearly six months (summer 2014) before it made national media and expressed frustration over what they deemed a generation, or two, lost to opioid addiction. As of the publication of this volume, this remains, for many, the top issue, and has not been adequately addressed. During the height of the COVID-19 epidemic, West Virginia led the nation in per capita deaths by overdose.[72] Respondents continue to point to a lack of adequate public or political leadership as the main issue of concern for their communities. This would concur with the theory regarding the effects of a conflated work/civic space on democratic practice and robustness.[73]

My questions, as approved by an Institutional Review Board in 2014, can be viewed in appendix C. Respondents more often engaged in similar discussions prior to my asking all of these; nevertheless, I ensured I asked this question of each respondent: what is your community's most pressing struggle?

At the time of writing this list of questions, I assumed that technology would provide a potential solution for part of what ails these communities. By contrast, respondents listed the following issues as top crises in their communities:

- Lack of leadership/Lack of Vision, which they equated.
- Opioid Addiction and Addiction Services.

With respect to lack of leadership/lack of vision, respondents explained (again, this is their view) that local politicians, including state politicians, shortsightedly voted or acted in favor of short-term personal gain, either in their acquiescence to the local single-sector industries, whether coal or natural gas, or, due to the prevalence and dominance of the energy sector, could not envision or explore economic opportunities beyond it. Though approving efforts to invent and encourage tourism, such as that of the Hatfield-McCoy All Terrain Vehicle (ATV) trail, respondents did not see enough economic support, training, or marketing to make that sector thrive.[74] If the region had better political leaders, and leaders with vision, then its myriad of other social, economic, environmental, and racial issues would not be as dire. Gaining significant ground on any of these, respondents found unlikely in the public or governmental sphere. Local and state politicians had no vision. This region was falling behind in population and thus, had fewer votes that mattered. These sections of West Virginia and Virginia were far from the state capitals—thus, out of sight and out of mind. Moreover, significant federal engagement was unlikely as this area just could not compete in terms of numbers served, and in federal granting and attention, quantitative measures of impact put the descending populations in the coalfield counties at a disadvantage. Federal grants were usually aimed at helping the most people. Rural areas needed assistance but could not provide the large numbers served per dollar spent.

Respondents expressed that local residents needed hope and that local political leaders lacked the ability to provide this either in inspiration or concrete achievement.

Respondents also indicated experience with current issues of political corruption, cronyism, and factionalism. They expressed dismay that often civic groups spent more time competing than collaborating. They pointed to direct run-ins with systems of kickback alive and well in city and county governments. They expressed concern, like that of Huey Perry more than fifty years prior, of misappropriation locally of federal funds.

In the cases of grasstops not working as public officials themselves, respondents described a need to work outside the local political system to achieve social change gains. They described their own community initiatives and those of faith-based groups as particularly significant.

Though not expressed in these terms, respondents prioritized the local democracy deficit of not having local political leaders acting on the true behalf of local private citizen constituents, leaders without vision beyond or in addition to the local single-sector industry, and leaders not being transparent or trustworthy as the major obstacles to addressing the wider range of economic, health, social, educational, and environmental issues. Without leadership, adequate advocacy to combat the region's issues could not occur.

And while this chapter focused on the specifics of an Appalachian coalfield and extractive spaces in Appalachia, I want to return to my original impetus for reflecting on a specific case as a kind of local explication of the phenomena more broadly in rural industrial areas constituting Big Rural. As the scholars I drew from and the interviews I conducted illustrate, the kinds of corruption and the kinds of challenges to the practice of democracy in an area with a dominating extractive industry are not concentrated in a certain country or geography. The first time I heard of anything conceptually like a "resource curse" came while I was working for a nongovernmental organization in 2009 in the formerly Soviet South Caucasus. Having traveled to Baku, Azerbaijan with my work, I was sitting on a hotel rooftop at a reception hosted by executives from what then was Statoil, the Norwegian state-majority owned oil and gas company. The NGO I was working for managed corporate social responsibility projects in youth and women development in the South Caucasus for both Statoil and British Petroleum (BP).

Baku was booming with natural gas and oil money and related new building construction. A local female democracy activist was also attending this event, and we got to chatting. The NGO I worked for also supported democracy-strengthening projects for investigative journalism and civil society in the Republic of Georgia and Armenia, but shied away from such work in the dictatorship of Azerbaijan, lest it be kicked out (which it was at risk of anyway, due to its sponsorship of democracy-building work in Georgia and Armenia). I told her about a recent proposal I coauthored for European Union support for strengthening the fourth estate in the Republic of Georgia. This prominent woman piped up—I wish we didn't have oil or natural gas, then we could be free here like they are in Georgia! She explained that the powers that be from Baku to Moscow would not permit democracy as Azerbaijan was too resource rich not to be controlled or exploited.

I pointed out that the Republic of Georgia also had its issues as the pipeline from Azerbaijan flowed through it to Turkey and on to the Mediterranean. She added that this pipeline increased the strategic importance of the region from a former Soviet backwater to a place now on the world stage of military and political strategy. To achieve this, people were thrown off their land to get this pipeline built. Environmental concerns tossed aside. No public comment or citizen commitments were sought; the entire project was orchestrated in the region with no civic actors participating.

I would trade all this, she gestured to the booming development, for freedom and democracy.

Although I didn't have the term "resource curse" at the time, I said, that because of where I was from, I knew something about what could happen if the boom slowed down or stopped.

She said they had been there before with the collapse of the Soviet Union. They could live through that again.

As off-shored spaces, as cogs in Big Rural, both Bluefield and Baku are in the crosshairs of global demands, corporate imperatives, backroom deals, and broken civil societies.

CONCLUSION

I have argued that poverty in the coalfield Appalachian space stems from layers of economic and social factors related to workplace automation, the shift toward larger-holding farms in the U.S. Midwest and West mid-twentieth century, unmitigated single-sector economies with no state or corporate policies of promoting transparency or democracy, and the totalitarian structure of the single-sector workplace and social space. No essential social, economic, or scientific law accounts for the strong correlation between a space with high fossil fuel production and low democratic function.

I have also claimed that the totalitarian nature of the workplace in the United States, with its top-down structure of command, extends into the social space in the rural industrial single-sector space. The inability to separate the social demands of the totalitarian workplace from one's social sphere contributes to the inability of a person living in this space to function as a democratic "free agent," able to make personal or political choices against the workplace imperatives. In the Pocahontas Coalfield, the union acted as a social organizing force and power counterbalance to the coal and land companies until the era of Reagan and increased union-busting or delegitimizing tactics of the 1980s.

The long-standing deficit in work in the Appalachian space and the historic fallout from severe cuts to a local workforce due to automation in the Pocahontas Coalfield by 1960 continues to reverberate in the region's long list of social, environmental, and health issues. Rather than listing these issues and employment as their top concerns for the Pocahontas Coalfield, qualitative interview respondents from among grasstops highlighted corruption and ineffective and uncommitted political leadership as their subregion's main concerns. In short, in the view of respondents, without transparent and ethical leadership with strong vision beyond the parameters of the historic single-sector industry, and without strong grasstops leadership, the region will remain short on resources it needs to address job creation, poverty, brownfields, and health-related issues stemming from the local single-sector economy. Chapter 2 addresses the founding of the Pocahontas Coalfield as a highly technical rural industrial space, with science and technology driving that space, and coproducing the culture there and beyond, especially the black boxing of large technical systems.

NOTES

1. Jerry Bruce Thomas, *An Appalachian Reawakening: West Virginia and the Perils of the New Machine Age, 1945–1972. Vol. 12* (Morgantown: West Virginia University Press, 2010).

2. Joseph Schumpeter, *Capitalism, Socialism and Democracy* (1947, Reprint; New York: Routledge, 2010).

3. Thomas, *Appalachian Reawakening*, 11.

4. Ibid, 13.

5. U.S. Department of Labor, Mine Safety and Health Administration, "History of Mine Safety and Health Legislation," n.d., accessed July 19, 2023, https://arlweb.msha.gov/MSHAINFO/MSHAINF2.htm; Thomas, *Appalachia Reawakening*, 15.

6. Howard Berkes, "Advanced Black Lung Cases Surge In Appalachia," *All Things Considered,* December 15, 2016, http://www.npr.org/2016/12/15/505577680/advanced-black-lung-cases-surge-in-appalachia; Thomas, *Appalachia Reawakening*, 15.

7. Thomas, *Appalachian Reawakening*, 15.

8. Thomas, *Appalachian Reawakening*, 287–92.

9. Defined more fully later in this chapter. Ibid, 293.

10. Ibid, 297.

11. Ibid, 299.

12. Ibid, 303.

13. Ibid, 25.

14. William P. Browne, *The Failure of National Rural Policy* (Washington, DC: Georgetown University Press, 2001).

15. Ibid.

16. Thomas, *Appalachian Reawakening*, 26.

17. Ibid, 28–39.

18. William R. Freudenburg, "Addictive Economies: Extractive Industries and Vulnerable Localities in a Changing World Economy," *Rural Sociology* 57 no. 3 (Fall 1992): 305–32.

19. In 2016 I brought this up to the person who does statistics for the Appalachian Regional Commission, and I have brought it up to that federal agency several times since—that we have no idea how many people ought to live in a certain area and what industries ought to support them. Although population decline has been devastating the region since the 1950s, no federal agency or policy analysis analyzes how best to approach it.

20. Edwin E. Holm Jr., *Production and Marketing of Coal in Virginia and the Nation* (Richmond: Virginia Division of Industrial Development and Planning, 1955).

21. David Leadbetter, "Single-Industry Resource Communities: 'Shrinking,' and the New Crisis of Hinterland Economic Development," in *The Future of Shrinking Cities: Problems, Patterns and Strategies of Urban Transformation in a Global Context*, ed. by Karina Pallagst, et al. (Berkeley: Center for Global Metropolitan Studies, Institute of Urban and Regional Development and the Shrinking Cities International Research Network, 2009), 89–100.

22. Jonathan Di John, "Is There Really a Resource Curse? A Critical Survey of Theory and Evidence," *Global Governance* 17, no. 2 (2011): 167–84.

23. The World Bank, Europe and Central Asia Unit. "Monotowns: Path to Resilience: A Brief Review of International Experiences with Urban Regeneration," 2010 (page discontinued).

24. Langdon Winner, *The Whale and the Reactor: A Search for Limits in an Age of High Technology* (Chicago: University of Chicago Press, 1986). The manipulation

or introduction of technologies in the production process to eliminate workers for reasons of politics rather than production efficiency has a long history. The case of the National Union of Iron Workers and Cyrus McCormick II illustrates this point. Being at odds with this union, McCormick II aimed the introduction of technology squarely at the most skilled workers, seeking their displacement through machines. Though the first technology introduced to replace these workers produced lower quality castings than those of the human workers, the machines achieved McCormick II's political ends of weakening the union.

25. Ibid.

26. Leadbetter, "Single-Industry Resource Communities," 2007.

27. Timothy Mitchell, *Carbon Democracy: Political Power in the Age of Oil* (New York: Verso, 2011).

28. Ibid.

29. Ibid.

30. Winner, *The Whale and the Reactor*, 1986; Mitchell, *Carbon Democracy*, 2011.

31. From this http://www.shrinksmart.eu/: Welcome to SHRINK SMART—*The Governance of Shrinkage within a European Context. Over the past years shrinkage has become a "normal pathway" of urban and regional development. All across Europe cities and regions have experienced economic downturns, out-migration and demographic imbalances and as a consequence "urban shrinkage" has become a main challenge for urban development.*

SHRINK SMART studies how this challenge is met by policies and governance systems in different types of shrinking urban regions. It is based on comparative case studies from seven urban regions throughout Europe. The project aims on analyzing different trajectories of shrinkage, understanding main challenges for urban planning and elaborating alternatives for urban governance.

As a result of the project, different trajectories of shrinkage processes within Europe will be defined and a set of policy recommendations for different constellations of shrinking cities will be developed. SHRINK SMART will engage in extensive dissemination activities and develop a set of tools and policy recommendations.

Shrinksmart is a collaborative research project funded by the 7th Framework Programme/ Socio-economic Sciences and Humanities of the EC.

The project was finished by 30 April 2012. The last update of this homepage was realized in December 2012.

Project title: SHRINK SMART—The Governance of Shrinkage within a European Context

Project number: 225193

Duration: May 2009–April 2012

Project budget (EU contribution): 1.496.092 Euro

32. The World Bank, "Monotowns," 2010.

33. Ibid.

34. Mitchell, *Carbon Democracy*, 2011.

35. Winner, *The Whale and the Reactor*, 1986; Stephen L. Fisher and Barbara Ellen Smith, editors, *Transforming Places: Lessons from Appalachia* (Champaign-Urbana:

University of Illinois Press, 2012). Fisher and Smith examine a range of practices that contribute to democracy beyond voting. These practices are worth review as they point to the deep poverty of focus only on democratic practice through voting and also carve out the territory for the case of the need for high and diverse civic engagement for a robust democracy to occur:

> Build bridges with allies. Show up at the events of allies.
> Share limited resources: money, time, and the spotlight.
> Advertise.
> "Get serious about racism and oppression." (27)
> Attend public hearings.
> Write fundraising appeal letters.
> Hold mass demonstrations and rallies.
> Lobby in Washington, DC.
> Conduct civil disobedience.
> Conduct a series of action campaigns.
> Hold vigils.
> Use the creative arts.
> Create puppet shows.
> Do street theater.
> Create art together in public around the issue.
> Do activism as a second job in addition to the job that pays the bills.
> Write letters to the editor.
> Create a "free" newspaper and print folks from your side in it. Distribute it widely.
> Get some local experts involved.
> Involve faith communities.
> Involve students.
> Involve young people.
> Build bridges among religious folks around key issues.
> Host events when the students are still in school.
> Use social media.
> Train volunteers to fundraise.
> Build upon each small success. (43)
> Provide background at meetings in order to welcome newcomers. (44)
> Have courage. (44)
> Multi-issue groups often garner more support yet single-issue groups often have a lot of energy (44).
> Form a nonprofit organization. (50)
> Highlight issues of livelihood and health.
> Be prepared to counter the messages that you are loony. Don't let them get you down. (52)
> Form citizen groups. (54)
> Get folks envisioning the future they want. (55)
> Utilize local opinion and action.
> Show local groups how joining a larger effort "will advance their own agenda." (56)
> Give people "principled reasons" to form alliances. (56)
> Welcome newcomers in the area to participate in civic life. (56)
> Reflect on whether joining larger networks would work to your advantage. It may not. (57)
> Create movements or actions based on values rather than interests—values last longer. Envision together the world/the community you want—this reflects values rather than issues. (57)

Use the principles of the 'solidarity economy': "reciprocity, democracy, sustainability, and equity." (59)
Help people know who they are and have a sense of identity. (65)
Start a festival. (69)
Conduct solid research.
Hold fundraisers such as turkey shoots, cakewalks.
Relate your mission to religion. Use Bible verses.
Create a listserv.
Write resolutions and bring them to city and town councils.
Ask allies for email addresses.
Be dogged.
Call people directly on the phone when social media or email does not get the response you need.
"Choose an initial fight you can win." (36)
Structure your group to be able to "quickly absorb newcomers." (37)
Create a steering committee to govern.
Create bylaws.
Reach across party lines.
Believe that "it does not matter if you have no chance of winning. It does not matter that no foundation will fund you. You fight viscerally because it is right to oppose something so awful." (38)
"Usually only a few respond with deep commitment, but often that is enough." (38)
Embrace democratic statements.
Table at local events.
Create a petition.
Offer cheaper and better environmental solutions.
Advocate a positive vision, but also attack the negative. (40)
Use volunteer labor.
Get a volunteer to do the website.
Hire a part-time person when you can afford it.

36. Andrew Feenberg and Michel Callon. *Between Reason and Experience: Essays in Technology and Modernity* (Cambridge, MA: The MIT Press, 2010).

37. Tim Kurz, Shona Crabb, and Martha Augoustinos. "Contesting the 'National Interest' and Maintaining 'Our Lifestyle': A Discursive Analysis of Political Rhetoric around Climate Change." *British Journal of Social Psychology* 49, no. 3 (n.d.): 601–25.

38. Daniel Breslau, "Economics Invents the Economy: Mathematics, Statistics, and Models in the Work of Irving Fisher and Wesley Mitchell," *Theory and Society* 32, no. 3 (2003), 379–411; Stirling, Andrew. *On the Economics and Analysis of Diversity* (Brighton: University of Sussex, Science Policy Research Unit, 1998).

39. Richard Ledet, "Correlates of Corruption," *Public Integrity* 13, no. 2 (2011): 149–62.

40. David Harvey, *A Brief History of Neoliberalism* (New York: Oxford University Press, 2007).

41. Winner, *The Whale and the Reactor.*

42. Richard Auty, "Natural Resources, Capital Accumulation and the Resource Curse," *Ecological Economics* 61, no. 4 (March 15, 2007): 627–34.

43. Di John, "Is There Really a Resource Curse?"

44. Deborah Stone, *Policy Paradox: The Art of Political Decision Making*. New York: Norton, 2001.

45. Foucault, Michel. Society, Territory, Population. Lectures at the Collège de France, 1977–1978. London: Picador, 2007; Dan Rodrik, Arvind Subramanian and Francesco Trebbi. "Institutions Rule: The Primacy of Institutions over Geography and Integration in Economic Development," *Journal of Economic Growth*, 9, no. 2 (June 2004): 131–65.

46. Rebecca Scott, *Removing Mountains: Extracting Nature and Identity in the Appalachian Coalfields*. Quadrant (Minneapolis: University of Minnesota Press, 2010).

47. Mark E. Warren, "What Does Corruption Mean in a Democracy?" *American Journal of Political Science*, 48, no. 2 (2010): 328–43.

48. Warren, Mark E. "What Does Corruption Mean in a Democracy?" American Journal of Political Science, 48, no. 2 (2010): 328–43.

49. Thomas Stafford, *Afflicting the Comfortable: Journalism and Politics in West Virginia* (Morgantown, WV: West Virginia University Press, 2005).

50. Mingo County remains one of the largest coal-producing counties in West Virginia.

51. Huey Perry, *"They'll Cut Off Your Project"*; *a Mingo County Chronicle* (New York: Praeger Publishers, 1972).

52. Thomas, *An Appalachian Reawakening*, 308.

53. Cynthia Duncan, *Worlds Apart: Why Poverty Persists in Rural America* (New Haven, CT: Yale University Press, 1999).

54. Larsen, Erling Roed. "Escaping the Resource Curse and the Dutch Disease?: When and Why Norway Caught Up with and Forged ahead of Its Neighbors." Statistics Norway, Research Department, 2004. http://hdl.handle.net/11250/180569.

55. Duncan, *Worlds Apart*; Daniel J. Graeber, "Statoil Completes Sale of Marcellus Acreage," *Home / Energy News*, September 16, 2016. https://www.upi.com/Statoil-completes-sale-of-Marcellus-acreage/1601474030319/.

56. Cowie, J., and J. Heathcott. "The Meanings of Deindustrialization." In Beyond the Ruins: The Meanings of Deindustrialization, 1–15, 306–7. Ithaca, NY: ILR Press, n.d.

57. Interview with respondent, 2014; Body of work by Christopher Barton: http://www2.ca.uky.edu/forestry/bartoncv.pdf

58. By contrast, only recently designated "Appalachia" by the Appalachian Regional Commission, the city of Youngstown, OH, for example, has reinvented itself from a steel town to a commuter town. Pallagst, Karina. "Shrinking Cities in the United States of America: Three Cases, Three Planning Stories," in *The Future of Shrinking Cities: Problems, Patterns and Strategies of Urban Transformation in a Global Context* (Berkeley, CA, February 2007), by the Center for Global Metropolitan Studies, 81–90. Berkeley, CA: Institute of Urban and Regional Development and the Shrinking Cities International Research Network, 2009.

59. Interview with qualitative interview respondent conducted by author in 2014.

60. The syllabi and programs at local community colleges, local colleges, high schools, and so forth reflect national trends or single sector interest.

61. US Census data.

62. US Census data.

63. Amanda Holpuch and Nadja Popvich. "Rural Counties across the US Becoming a Powder Keg for HIV Outbreak." *The Guardian*. July 31, 2016.

64. Ibid; Centers for Disease Control.

65. Shannon Bell, "'There Ain't No Bond in Town Like There Used to Be': The Destruction of Social Capital in the West Virginia Coal Fields." *Sociological Forum* 24, no. 3 (September 2009): 631–57.

66. US Census data.

67. Some examples: Barbara Ellen Smith, "De-Gradations of Whiteness: Appalachia and the Complexities of Race." *Journal of Appalachian Studies* 10, no. 1/2 (Spring–Autumn 2004): 38–57; Joe Trotter, *Coal, Class, and Color: Blacks in Southern West Virginia, 1915–32*, Blacks in the New World (Chicago: University of Illinois Press, 1990); Deborah Weiner, *Coalfield Jews: An Appalachian History* (Chicago: University of Illinois Press, 2006).

68. Jessica Contrera, "In a Place of Need, an Unhealthy Contradiction: They Are Poor, Sick and Voted for Trump. What Will Happen to Them without Obamacare?" *The Washington Post*. March 11, 2017.

69. Thomas, *An Appalachian Reawakening*.

70. Christopher Barton (Professor of Forest Hydrology and Watershed Management at the University of Kentucky) in discussion with author, July 2016. Much land has chemical contamination and/or soil fertility is prohibitive.

71. Grasstops are leaders of nonprofits or other groups of decision-makers.

72. West Virginia Department of Health and Human Resources, "West Virginia Experiences Increase in Overdose Deaths; Health Officials Emphasize Resources," *2021 News and Announcements*, April 23, 2021, accessed, June, 6, 2022: https://dhhr.wv.gov/News/2021/Pages/West-Virginia-Experiences-Increase-in-Overdose-Deaths;-Health-Officials-Emphasize-Resources.aspx

73. Ledet, "Correlates of Corruption"; Winner, *The Whale and the Reactor*.

74. The Hatfield-McCoy Trail: http://www.trailsheaven.com/. http://www.trailsheaven.com/. "The Hatfield-McCoy Trail System is a statutory corporation created by the West Virginia Legislature to generate economic development through tourism in nine southern West Virginia counties."

Interlude 1

Jason Tartt, Sr.

I have always conceived of *Big Rural* as not only expository or theoretical, but also as witnessing, and, as filling in the gap in practice I was seeking guidance in, and when attempting on-the-ground work of making any difference on the issues current in the Pocahontas Coalfield.

I am also acutely aware that my theoretical work does not occur in a vacuum, and that I did not only want to engage with the community I am from as a further abettor to extraction, in this case, knowledge extraction for academic inquiry.

Research fatigue in the rural is real, with academics and media makers wanting to extract stories and ideas, and then often, never reporting anything back in return, not even bothering to show up to present on understandings or creative inspiration the folks in the rural may have lent them.

The same is also true in the nonprofit or social change arena—often folks in other less extractive sections of central Appalachia assume the localized experience of people in the Pocahontas Coalfield translates to their locale, and, they can't be bothered with actually visiting there. They write off the words of coalfield folks and their explanations of the issues at play as excuses (again, the deficit model I bring up in the *Preface*—assumptions that people in Appalachian coalfield areas can't be trusted to know what to do, to know what is at stake, to know what is wrong, or to know what is right; they cannot be complex; they can't be allowed agency).

The three interludes in this volume serve to open space for those who have taken on *doing something* about the state of the Pocahontas Coalfield. Together in dialogue we form a quadrant of people born in and affiliated with four interconnected coalfield counties, from east to west, Mercer and McDowell in West Virginia, and Tazewell and Buchanan in Virginia. I will pick up more on the thread of the importance of this work, and reflect more on this process in Interludes 2 and 3, and the Epilogue.

I place each interlude as an auxiliary to the academic material prior to and following it. Jason Tartt, Sr. brings a name, face, and family to the Pocahontas Coalfield experience of the Black community in Vallscreek, West Virginia. He walks us through the Deep South and Up North connections, and, through his current work reigniting mountain farming, and, reigniting Black excellence.

Jason Tartt, Sr. (he, him, his).

During this interview, Jason was speaking to me from Ghana, West Africa, where he has found he has ancestral roots, and where he also does development work, investment, and personal healing.

Jason: I have been doing some interesting reading this trip, a Joe Trotter book about Black coal miners. The Trotter family used to live in Vallscreek, WV [where Jason lives and much of his family also lives].

Getting into this book by Joe William Trotter really gave me the background on the migration that took place in the early nineteen hundreds. This is very personal for me, not just from an African American standpoint, but just from an Appalachian Mountain West Virginia standpoint.[1]

I'm sitting here in Ghana right now, where my ancestors were from. You know the migration from the South into West Virginia, and it's the same thing. And you see the remnants of what the British have left here, which was nothing. They left the place for dead, and it's been a constant struggle ever since. But to see the resilience, to see the "no quit." People struggle with a smile on their faces making something out of whatever they have is the same thing that I grew up seeing and somewhere along the lines, I think that was lost.

That's what took me by surprise when I got back to West Virginia (after living in other places outside the region)—what in the hell happened? Not just to the place, but to the people, to the fight that these people used to possess. It was like everybody gave up, you know, and that's just not the place that I remember growing up, White. Black. Otherwise. These people were resilient as hell. They were resourceful. They made something happen.

And there's this stigma with the hillbilly and all that, but turns out some of the folks that I know here turned out to be some of the most brilliant people that I've ever met. This makes you reassess. What is intelligence?

I'm proud of where I'm from. I'm proud of the people that are there. I feel obligated in ways to do what I can to bring a sense of pride and ownership back to the community there. Coming from the African American community, it's well known that whatever struggle we've endured, what we've dealt with was not by accident.

I remember being eight years old and looking up at my dad standing there, near Vallscreek, in the middle of the road, looking up at my dad and asking him why us? How come we have to be poor? Why don't we have what other people have? Why do I have to be called "the nigger" when I go to school? The truth of the matter is it's the same game and the same thing that was done to us as I can see being done to people in Africa. And as Martin Luther King, Jr. said, an injustice anywhere is a threat to justice everywhere.

At some point, we have to stop this nonsense, and we've got to get it in the fight. Otherwise, what's already happened, my children can look forward to, and my grandchildren can look forward to the same happening to them. And you've got these people in this community that dosn't know anything other than what their dad and their granddad's told them. When you sit down and have a conversation with them, and you tell them those days are over, even if they do bring coal back, how does that translate to jobs? How does that affect this community? I've always told my own children—if you want to see change, then you have to roll your sleeves up and be a part of the change that you want to see, and that's it for me.

Crystal: Where does that come from for you?

Jason: As a Black man a couple of thoughts come to my mind. Something that was drilled into me a long time ago is you never forget where you come from, and so many times you see people from the Black community that make it, which get the money that find a way out, they forget where they came from. They forget that struggle. They forget those who helped them get to where they are, and so that was always instilled in me, and that stuck with me.

Another thing is, I come from some very strong people, and my great-grandmother, her name was Alberta Jackson. She helped raise me. This woman was raised just shortly after slavery ended. Not fifty years had gone by. And what I learned from her is to never quit. I don't care if it's killing me.

I remember my grandfather, her son, told me a story. He was working for a White man in Alabama. This White man left some money lying around. My grandfather saw it and gave it to the man's wife and then left. The White guy comes back, and then keeps in mind, this is 1920 in Alabama. This dude grabs a posse and goes off to my grandfather's house. My great-grandmother, the one that helped raise me, was there, and they came and accused my granddad, and he said I gave the money to your wife at your house. They say, you better hope you did, or else you'll probably be swinging from a tree here in a little while. That's just what it was, so they grab him up. My grandmother had no power in anything but to let them take her son. They find the wife and she says, yeah, he gave it to me. When you hear things like that, you know, when you hear what your ancestors went through—When you come to Ghana and you go through these slave dungeons, you realize that you are standing on the shoulders of your ancestors, and without these people being strong enough to endure what they endured, and to go through—what they went through for me to be sitting here talking to you today, then I can never give up. I'll never quit.

I have to draw on memories of growing up and seeing what's been done and realizing that what's happening in McDowell County has happened before. If we're not going to stand in front of this thing and fight for this community and for these people, then nobody is going to do it.

I realized that it could be worse. Being in places like West Africa right now, and seeing a much worse situation. Having been in the military, and being in places like the former Yugoslavia in Bosnia, what's called Bosnia now, and to have lived through that, to make it through that.

You look at countries like Rwanda, and I've got friends in Rwanda right now. They went through genocide less than forty years ago, and now it's one of the cleanest, fastest growing economies in the world. Those are some of the reasons why I believe that it can happen in McDowell County. I feel like I'm equipped for it with my background. Everything that I've gone through from my corporate experience, military experience, sports, being in a community of mixed race and multicultural cultures.

Number One is to generate a sense of pride, and remind people of where they are. I think we all know how to build sustainable communities: it starts on the ground. It starts with education and building this thing in a way that the community can benefit from. No one has come here and spoken in this language before, because it's all been focused in the hands of a few people.

This time we want the community to be on the front lines and have a share or a stake in what's going on. Most people that come, they're surveying and studying and feasibility studying, and they put information together, and they leave. Several of my closest friends here have died from drug overdoses. When you see eight-year-old kids in your community walking around with full-blown diabetes. Then you have grandchildren of your own. You have children of your own. You want something different, better, for them.

Empathy is something that I believe strongly in, and I believe in putting myself in other people's shoes. What if that were my child? What if that were my mom and dad who's an elderly couple who doesn't have any help? I draw on empathy, that those could be my children. You can't say you're you about your community if you're standing by and not trying to do anything to help that situation. And this certainly is my community.

When I came back to West Virginia, I would ask people, where is so and so? So and so is dead. What about so and so: dead. There are so many people that I went to school with, people that I was friends with that have died. There was a twenty-four-year-old young girl just last week—this kid was twenty-four years old and died of a drug overdose.

When I first got started with agriculture here and wanted to make a case for it to be part of what is studied in the schools we would go to the high schools here, and those students would say, Well, why don't get the same opportunities that kids that are in other places get? Or you ask those kids, or what are you going to do when you graduate? They answer; get the hell out of here.

Crystal: What we're really asking for is for McDowell County, for people to live there like anywhere else, right? We are not asking for special treatment.

Jason: When we moved back to West Virginia, my mom was the person who actually got me into gardening. I had done some substitute teaching here in McDowell County and I moved back, but I knew that wasn't going to work for me just simply by the way that the system was being managed, and how the children are being taught.

Agriculture made a lot of sense. I'd started buying land. I didn't know what I was doing, of course, didn't know a thing about farming. Now, we've been at this ten years, but I think probably five or six years in, I really got serious, no more of this nonsense. I had to sort out the real potential. The first step in that was working with Appalachian Beekeeping Collective [a Hinton, WV based nonprofit promoting beekeeping as a regional industry] and getting into these beekeeping classes.

All those other years we were trying to fit into what they were telling us, trying to go along with their programming. We were used in photo ops with the new Ag commissioner. They just want to come and use us as a couple of tokens.

Crystal: They have a federal mandate. If they're going to get federal dollars, they have to show that they have done XYZ and show diversity. You played their game for a while. But what about now?

Jason: I think part of it is calling people out. I gave a lecture at WVU a few weeks ago, and the Dean of Eberly College was there. I did a little looking into Land Grant universities before I did this talk because they're a Land Grant University. Land Grant universities were established during a time of real change in the United States, and I let them know we're there again.

I said we need you folks to step up. We can't do it without you. I said you're complicit. When you look at the opportunities that exist with respect to agriculture in our region you've got a major university like WVU sitting there. You're supposed to be a land-grant university. And WV State University is one and also an Historically Black College or University. And here you all are still talking about homesteading.

Let's build education and awareness around the real potential for agriculture and build infrastructure around this. Anything that's going to be sustainable has to come with education. If you're not talking about education, it's not sustainable. After we have this lecture, and we sit down and talk, it's not enough for you to tell me what's wrong. What do you want to do? I want to know what you are going to do to contribute to the solutions that we need, to show the capacity of what Appalachia could be.

Agriculturally speaking, no one's done that research. If you want investment in here, then you've got to give data. You've got to have information to show people, and who better to do that than WVU? We understand you are now like a corporation. We understand that you need to make money with these things. So, let's figure out how to get the money in your coffers so that we can get the data, the research done so that we can show investors and get money in here because it's going to take money to drive these changes.

I have been working on getting in front of the right people with this. I did another talk in Tampa, FL with an organization that flew me down there, and they were doing some Farm Bill work. They started talking about the Farm Bill. I said, let me tell you what I think about the Farm Bill and the rest of your bills. They are very short sighted. You look at the Farm to School program, and no farmer in Southern West Virginia can offer any school what's on these lists that you all come up with because that's not who we are. We don't produce those products, and we don't have the capacity to produce those products in the volumes that you need.

Now, if you want to talk about a real Farm to School bill, it would include Appalachia with its own footprint of what it has to offer. We need to put some real thought and some real effort behind creating these bills so that they make sense, and that they're inclusive of all regions across the country. They need to include: what could Appalachia offer?

Obviously, we could be massive fruit producers, and this would have not just an impact in our region, but it would have a national, if not international, impact on the amount of fruit that could be produced throughout Appalachia.

You talk about honey. You talk about maple, you talk about medicinal herbs, which is a huge industry. There are so many things that we should be discussing, and people are like, why didn't we know this or think of this?

Crystal: Because you didn't get out and ask.

Jason: More than not getting out and asking, they were getting money from being a Big Ag model. With that, you are missing out on a lot of opportunities that can affect the economy, health issues, jobs, and small business creation.

You could be known for ending poverty and food deserts. We need you in the fight with us to do this. All of this abandoned mine land that is contaminated, there are even opportunities there. I mean, you could use agriculture to fix those problems and grow products like bamboo or lavender.

Also, everybody's talking about agriculture, agriculture, but nobody owns any land in McDowell County because it's owned by probably four or five corporations. We sat down and built a relationship with those people who owned the land. It took about three years for us to finally sign the deal.

People's ability in this region to make a living has always been tied to the land here, and that's been taken away, so we need to get land. We don't need it all. I'm not here to argue about coal mining. I'm not here to do that at all. There's room for everybody to make money if coal mining is doing great, fine. But we're more than coal.

That's been the biggest point of discussion for me with these landholding companies is just that we're more than coal. We're more than timber. There's a bunch of things that we could be doing if you need help fixing some of this contamination. Agriculture can provide solutions for that. When you step back, and you look at just the cleaning up the mine land by itself, if we didn't do anything else, and we used agriculture to clean up some of this mess that they've left behind, that's a huge market right there, and then you start talking about honey and all the other products that come from beekeeping, maple, medicinals, and fruit.

I took some folks up on a mountain one day and as long as we were driving you saw blackberries on both sides of the road for miles. Then you've got a sixteen year-old in our training cohort. He has sixteen acres over in Bradshaw, and he's sitting there like he's got something on his mind. I said, what's up? He says, you know, you really got me thinking hard about these blackberries. That gives hope to our cohort members, because when you frame it the right way for these people. He is thinking of his sixteen acres in blackberries.

Crystal: I wanted to circle back around to the beginning of our conversation when you said you do this work also out of a place of empathy.

Jason: You know, community building was not something that I had any experience doing. In a place like McDowell County, most people would not do what we're doing with the folks we work with on a daily basis. It took us three months just to get the cohort, the positions filled.

Come payday on Friday there is a likely chance that two or three, if not more, are not going to show up on Monday, so you have to understand what you're getting into. Understand what this is, and it's not necessarily anyone's fault, but that it's a process. It's a process for us to see someone struggling with methamphetamine when they start, then not doing meth at all for a month. Four months or so in with us, coming to work bright-eyed and willing to learn, ready to work, asking questions, and really all in with what we're talking about.

My dad and I were sitting down, having a conversation before we started this last training cohort. I said this is not one of those situations that , people are going to show up to work. He says, get it done. He said, you're going to have to be very patient. You know who you're dealing with. You know this community. You're going to have to make it work. I have had people say, I wouldn't put up with that. I know you wouldn't. But that's why you're not. You're not here doing this work.

Empathy, there again, and really make sure that you're cut out for this work because it's not for everyone. And if you're not used to dealing with people who have checked out on life . . . by the time we're in the second or third month, you're starting to see change. You're starting to see people gain more interest.

You have a few that may go in with the right attitude from the beginning, and when others see their progress and what their mindset is and we're bringing folks in from other organizations like Mark Lily [master beekeeper, Appalachian Beekeeping Collective] and West Virginia State University [doing a weather balloon project], that adds credibility to what we're doing. When these folks are seeing us at WVU, they're cheering us on. They're there. They feel like they're a part of this movement.

We're in the business of helping people, and that goes far beyond just economics, far beyond agriculture. It's people with real problems, who feel like no one cares. People who feel like they've been given up on, and really being able to dig into that, give yourself to that. That's been my whole life, because in the Black community, that's what it's always been, and you know people start to talk to you. I've had cohort members cry on my shoulder.

Now I remember the first cohort what we did, that particular cohort was when George Floyd was killed, and in that particular cohort we had nine or ten people, and of those nine or ten people, I'd say more than half of those people were young Black males. I remember being at work with them on the farm one day, and we were all standing around talking about it after it happened, and they were so full of anger, they didn't know what to do with it, plus the fact that some of them came out of prison. Some of these guys were from the streets, to begin with.

I knew that this could be a problem, and they called me Pop, from listening to my son. So, they said, Pop, what do we do? I felt like crying over it myself. I feel upset: when they're protesting, and they're doing all this work. And I look at these guys. One day we were out on my farm, and I said, Guys, this is our protest. This is our protest. They were all looking around and they got it.

You know the stereotype, the narrative that's been painted. The reason that Black men die is because this country has made a narrative around Black men for the longest time—that our lives aren't worth much: we're fit for prison, sports, and entertaining people. Outside of that, we're good for nothing. So, do something with yourself. Come out here and do this work. Let's get this job done. Let's get this thing done. Let that be your protest. Let's change the narrative here.

My goal is to see some of these folks who have come through our training realize their visions, and we're not creating visions for these people. We're just giving them exposure to the possibilities and allowing them to choose. And once they've chosen, we try to support them as best we can to realize that vision.

NOTE

1. Joe Trotter, *Coal, Class, and Color*: *Blacks in Southern West Virginia, 1915–32*. Blacks in the New World (Chicago: University of Illinois Press, 1990).

Chapter 2

Scientific Promises and Prosperity

Constructing the Rural Industrial Space

OVERVIEW

Science and technology enable the Pocahontas Coalfield to occur as a single-sector economy. From this particular socio-technical intervention and its details I draw lines to it as a type of Big Rural. The stakeholders in the Pocahontas Coalfield—private, state, federal, coal boss, politician, worker—construct that space as rural industrial. Technical knowledge opened the opportunity for industrial intervention and continues to construct coal mining and its accompanying coalfields as social and technical practices. In this, scientists are not absolved from the fruit their "knowledge creation" bears, whether those fruits be political, economic, social, or environmental.[1]

If science creates knowledge, then the large machines, in the case of surface mining, for example, not only destroy an environment, or contribute to work displacement, but also destroy a kind of knowledge, an intimacy of and practice of place. The Pocahontas Coalfield, its science and technology, and its inhabitants, as parts of a large technical system aka energy production become rendered invisible in the larger culture as extensions in the invisibility of ubiquitous technological systems such as energy or water provided to metropole or nonproducing (white collar) spaces (the ubiquitous invisibility of Big Rural). Science and technical practices vis-à-vis corporate entities can render a rural industrial space and its residents a kind of living dead, unable to democratically change their situation, stripped of rights of actionable protest, and economically, physically, or due to identity rooted in work, unable to leave. Residents in this space can become lesser "citizens" in terms of their abilities to construct and enable democracy and democratic practice, not just in the Pocahontas Coalfield, but in places rendered as its type.

Other academic work focuses on the less than forthright means by which the initial landholding companies came to possess the land in what became the Pocahontas Coalfield.[2] I take several steps back further into the past. Prior to even deciding to wrangle land from the locals, people with money to invest had to decide that these proposed coalfields were going to pay off and be worth the hassle of that pay-off.

At the onset of the Pocahontas Coalfield, the myriad of social and economic problems currently at play could not have been fathomed by its first and strongest proponents. Among these, Jedediah Hotchkiss, a geologist, and also a mapmaker and topographer during the Civil War for Stonewall Jackson, advocated strongly that Virginia ought to abandon its agrarian economic reliance and focus instead on its mineral wealth. What I call the "Hotchkiss' wager" was his insistence that this large technical intervention would bring prosperity if only capital and the state would invest. Hotchkiss insisted that focusing on mineral wealth instead of continued agrarian production would bring greater prosperity to more people in the two Virginias. To this end, he created and promoted a scientific journal to gain investment for this wager.

Evidence exists of Jedediah Hotchkiss' journal making the rounds among investors in the region in 1880, thus predating some evidence of other scientific assessments as the first to have broader investment influence.[3] Though the following represents part of this picture, in fact, many pieces had been in place and moving with respect to mineral exploitation in that region well prior to the Civil War.

The economic, social, personal, health, and environmental degradation of this region due to the culture and corruption of the single-sector energy sector in the Pocahontas Coalfield and other southwest and West Virginia coalfields as outlined in this volume could not have been what Jedediah Hotchkiss envisioned when he advocated for industrial scale coal production. With technological and technical advancement, totalitarian workplace and living space conditions, externalization of corporate environmental costs, and political corruption, the Pocahontas Coalfield morphed from prosperity to a Pandora's Box.

Coal country politicking, cronyism, and corruption tell one part of this story. Hidden in plain view (scholar Jerry Bruce Thomas addresses some of the technological changes directly in this "new machine age"), are other kinds of leaders—those whose expertise the industry relies upon.

The hard sell outside the "Virginias" for investing and how science and technology were used to frame this sell set up a narrative of industrialization, ties to the Northeast, technological sophistication, and worldliness, that, at the time, competed with the popular and likewise manufactured narratives of the greater and adjacent Appalachian region as being inhabited by backwoods, salt-of-the-earth "yesterday's people." This narrative of science, technology, and industry gets lost now in popular culture and in larger cultural associations of Appalachia.

Cultural narratives of rural spaces elsewhere also mask the more salient and aggressive impact of industry. In the rural space, we think of miner, farmer, or angler rather than mining engineer, agrochemical engineer, or biological systems engineer.

Because we fail to grasp the impact of scientists and engineers in the rural space or examine how the rural space is also categorized, classified, intervened in, manipulated, and with sophisticated technology applied to it, we lose sight of the full complex technical reality of the rural space. Let me state this again and more plainly for the Rural sociology, Rural Studies, and Appalachian Studies readers of this text: science and engineering determine the culture of highly technologically intervened rural spaces like coalfields as much, if not more, than other cultural productions. Serious rural and Appalachian scholarly inquiry ought to also follow the culture of science and engineering in these environments. If I could attach a flashing light to this passage I would because so much opportunity for theoretical reflection, concrete social change, and paradigm-shifting analyses becomes lost because of field disciplinarity or reification of liberal arts-associated *cultural* productions over the arguably fiercer impacts of the as much *cultural* productions of scientists and engineers.

This lack of inquiry results in scholarly, policy, cultural, and political consequences. Rather than science and technology for the benefit of people in the rural space and for land stewardship, we arrive at single-sector economies in a democratic deficit, with local populations and environments suffering. No, I am not accusing cultural or social studies areas of creating the socio-technical circumstances of the Pocahontas Coalfield, or, of Big Rural, but I am hurling "j'accuse" if once aware of the potential for the cultural study of what is impacting folks in the rural the most, and one professes to do scholarly work toward "social change," then one continues to black box the culture of scientists and engineers and "industry" in the region and to black box "capital." More than enough material, opportunities for observation, potential partners for inquiry, and archival material are available to fill an entire library with scholarship on science and technology/scientists/engineers intervening in or being used to intervene in the rural space, their arbiters and abettors, and the past, present, and future impacts/results of their collective labor, culture, outputs, inventions, investments, and identities.

Promoting Science and Data to Sell Nature and a Patriotic Plan for "Improvement"

Nineteenth-century industrialism was not without its social critics (Karl Marx) and even its environmental critics (Henry David Thoreau, John

Ruskin, Alexander Humboldt).[4] Accidents, injuries, fires, and greed were common associations with coalfields in the United States by the time of the opening of the Pocahontas Coalfield. For example, in his close examination of the Pennsylvania anthracite coal basin, Anthony F. C. Wallace surmises that in that area, prosperity had been fleeting due to coal operators ignoring scientific evidence about the viability of the coal seams (depth and breadth of accessible coal). Mining operators ignored issues of safety, not enough coal was produced to cover costs, technological advances did not deliver expected seam access, and despite the popular cultural displacement at that time of the war hero with the industrialist, the latter did not always rise to the occasion. In the case of the anthracite coalfields of Pennsylvania, the issues with production there were well-enough known: the coal seam was not deep enough, mine construction was poor, in the early years there were no safety regulations, miners were blamed for accidents, miners got "miner's asthma" from breathing coal dust, stone dust, powder smoke, methane, and carbon-monoxide gas.[5]

One can assume that enthusiasm for the development of the bituminous (a softer coal more suited to heating and a wider range of uses than anthracite coal) Pocahontas Coalfield had to be contextualized within the personal, financial, and community risks affiliated with the previously opened coal-fields in other regions.[6] The scientific assumption by Hotchkiss was that the Pocahontas Coalfield seam was wider, and the coal of higher quality, especially for metallurgical and energy production. Where science had been undermined in the St. Clair region of Pennsylvania by industrialists willing to mine anthracite coal despite large safety and financial risks—in the end those coalfields did not produce adequate returns. By contrast, Hotchkiss promoted science as a cornerstone of this new potential coalfield and as a necessary driver for sound investment.

Infamous to Civil War scholars or enthusiasts for some of his feats, yet unknown or little examined with respect to his scientific and economic advocacy, Jedediah Hotchkiss included successful lobbying for mineral wealth extraction in Southern West Virginia and Southwest Virginia among his most successful acts of "public service." Beyond his pivotal military role and his fluency as a professional scientist, regarding Hotchkiss, we must set aside current assumptions, and instead engage with nineteenth-century actualities pertaining to American science and technology and what qualified one then as a scientist or engineer. Hotchkiss' is a story of understanding the changing direction of the U.S. economy and the post–Civil War South as well as the role, up for debate then, that science and technology ought to play in building the economy. Though a respected surveyor and geologist (and deep friends with the founder of MIT, a fellow mining engineer and geologist), Hotchkiss did not undergo formal education as a topographer, surveyor, or geologist.[7] Moreover, Hotchkiss' tale as a businessman and technologist rests on competing narratives for the "two Virginias" (West Virginia and Virginia) and what their modern identities ought to be.

Hotchkiss created a scientific and technical journal as a promotional vehicle for attracting investment in the industries, which he envisioned as pivotal for bringing the two Virginias prosperity and himself and other mining engineers and former Confederates wealth, in returning him to his Civil War era glory, and for pulling the South out of a still agriculturally oriented dark ages. Hotchkiss did not act alone in his role as chief data, science, and industrial two-Virginias propagandist. Other self-taught or university-educated surveyors, geologists, and mining engineers often cum businessmen, land speculators, financiers, or coal "barons" worked with him or in his consort.

A Fight for "Development"

In addition to narratives of democratic progress tightly wound around the American project with respect to its Enlightenment roots of liberty and freedom, adjacent to this narrative also runs an American project of land grabs and land speculation and fulfillment of Manifest Destiny. Without resorting to presenting the obvious regarding European colonialist and White settlers' intentions and actions with respect to the acquisition of land in North America, narratives of progress, industry, utility, commerce, access, conflict, and scheming regarding land in North America go hand in hand with the advent of "America" as a place created. In the American context, the land was not only pecuniary but also mixed with issues of national or individual state interest and economic development.[8] For example, not only was George Washington involved in the formation of the United States as a democracy project, but he was also a major land speculator, and through this speculation, arguably, America's wealthiest sitting president. Washington's financial interest in land merged with his interest in the proposed utility of the land, and with his projections for what the American colonies, and later, the United States, ought to concern themselves/itself with economically.[9]

Though potentially difficult for us now to consider in the context of mature industrial capitalism, the investment capital and stock boom, and the armed labor struggles of the late nineteenth and early twentieth centuries, the years of Reconstruction concurred with the U.S. federal government contending with crushing national debt, the ruined economy of most of the former Confederacy, and how or if to convert the former enslavement-tied agricultural base of the South toward industrial production and its "free labor," and, thereby, to different economic opportunity and potential stability, if not prosperity.[10]

At this time, both agricultural and mineral extractive enterprises relied heavily upon human labor. In the South, both carpetbaggers and local Southern "patriots" sought opportunities to industrialize the economically "regressive" South. To the burdens and trials of industrialism, add the details of converting the former Confederate states back to the Union, immense war debt, ruined infrastructure, interpersonal and state hostilities, and shifts in Black and White political power.

A major interest at this time was in raising capital to service debt, rebuilding railroads, and in "modernizing" the former Southern enslavement-dependent economies through attracting outside free labor and investing in industries beyond agriculture.[11] Toward the latter, data and facts were needed to attract investment. Coal was a known quantity in Virginia as in the late eighteenth century an early bituminous coal industry had sprung up near Richmond that in the subsequent sixty or so years was eclipsed by investment and technological adaptation by the Pennsylvanian Anthracite Coalfields.[12]

In *Old Dominion, Industrial Commonwealth: Coal, Politics, and Economy in Antebellum America*, historian Sean Patrick Adams contends that state investment in scientific data gathering, technical infrastructure, and business policies may appear mundane each in turn, but that in the aggregate, they result in necessary and foundational developments for supporting the growth and competitiveness of emerging industries. In my participant observation in industry, my surveying of recent federal requests for proposals and of federal policy and strategy documents, and my engagement with scholarship on the impact of scientific and technological R&D on the materiality of our lives, also in the rural, these three "buckets" of modern life, scientific and technical research, technical infrastructure, and business policies, remain as salient now as they did pre–Civil War. And, they are as much culturally and socially created and maintained (if not more so due to the resources devoted to them) as other "cultural" productions, and, also, shape the potential of nearly every facet of our daily lives.

With respect to Virginia, until after the Civil War, politics served and protected large-scale enslaving landowners. Patrick likewise contends that not just the availability of enslaved people impacted industrial pursuits; enslavement also impacted the economic perspective—large enslavers dictated the parameters of exploitation—focusing more on agriculture than on other kinds of industry. State legislators responded to potential economic development projects both in concert and individually not necessarily rationally nor armed with facts. Prior to the Civil War, agriculture remained privileged in Virginia, and coal and other mineral resources remained not fully surveyed nor fully promoted till well after that war.[13]

Post–Civil War, a large concern of prominent Virginians was revitalizing its economy. To these ends, former Confederate Brigadier General John Imboden was involved in the economic rebuilding of Virginia, addressing issues of labor and land. As the demand for coal increased, with the demand for iron as part of broader industrial and railroad expansion, Imboden characterized coal as Virginia's highest prospect for return on investment. Directly after the War, Imboden explained this to a friend, "in short time digging a mint of money out of . . . hills and mountains. I know of no such opening in Virginia just now for big returns on a comparatively small capital."[14]

In 1872, Imboden detailed what was known regarding Virginia's coal and iron deposits, and the next year visited London to raise British capital and investment for developing mining in southwest Virginia. Imboden authored a paper considering the commercial potential for these minerals, which he read to the Virginia legislature in February of that year—the same month the U.S. Congress received a report on the status of other Southern states during Reconstruction, with a special examination of White on Black violence.[15] By 1875, with British investment and family interests secured in developing coal in Virginia, Imboden became the leading authority on mineral deposits in southwestern Virginia.[16]

In 1880, Imboden relocated to Abingdon, Virginia, and began promoting the coal stores of Dickinson, Russell, Washington, and Wise Counties. He employed a government geologist, Professor Stevenson, to assess the coal and iron deposits in his extensive land holdings there.[17] In 1886 for the U.S. Bureau of Statistics Imboden created a 300-page report on Virginia's mineral potential containing detailed site surveys and statistical analysis.[18] The U.S. Congress subsequently requested 25,000 copies of the report.

By 1888 Imboden was installed as general manager of the Mineral Bureau of Southwest Virginia, Eastern Tennessee, and Western North Carolina to promote that region's natural resources. However, the booming iron mining he envisioned for the region stopped short when the deposits revealed themselves shallow and the new "Damascus" Imboden envisioned building in Washington County, Virginia caved during the severe economic downturn of 1893, when "15,000 businesses failed . . . 150 banks closed . . . and 4,000,000 workers" lost their jobs in the already struggling United States.[19]

While Imboden died before coal extraction for southwest Virginia was realized, his associate Jedediah Hotchkiss succeeded. Both Imboden and Hotchkiss considered tragic for the State of Virginia the pulling of funding from the 1842 geological survey under the direction of geologist and scientist William Barton Rogers, though Rogers did complete six volumes of "Reports of the Geological Survey of the State of Virginia."[20]

In the early nineteenth century, geology was a nascent and largely incoherent field, with many amateurs focused on proving biblical provenance through land formation. Pennsylvania, with its booming coal industry prior to the Civil War, also had the first lauded American Geological Society. Until the rise of William Barton's brother, Henry Darwin Rogers, the National

Academy of Sciences viewed geologists as mostly under skilled and unscientific. Henry had been tapped to undertake a major geological survey of Pennsylvania and was also elected to the Geological Society of London. State support in Pennsylvania was granted.[21]

Prior to the Civil War, in Virginia, a stream of out-migration due to limited opportunities for non or small landholders led the state to explore possible solutions for stopping its hemorrhaging of White residents. A mineral survey was one proposal made to jumpstart an industry that might entice able-bodied White men to stay. Two years after a Virginia Geological Survey was proposed, it was begun by Henry's brother in 1835.

Like the Pennsylvania survey of his brother's before it, William Barton Rogers carefully outlined the economic benefits of the survey over its scientific aims. Both Henry and William, however, were strained by the political demands placed upon their scientific and knowledge-seeking endeavors. Due to pressure from agricultural interests in Virginia, William's survey stressed the mineral potential for agricultural fertilizer, causing western Virginia's mineral potential to remain underanalyzed and underreported. As geology was also an emerging science, William Barton Rogers struggled to find assistants whose methods and knowledge he could trust, and those whom he could trust found the terrain and conditions in the western section of the Virginia survey formidable. William also encountered legislative opposition regarding the practical applications of his reports. After 1842, the geological survey of Virginia was defunded, with its mainly western mineral deposits still only haphazardly revealed. By contrast, Henry succeeded in publishing his reports on Pennsylvania, and, the Pennsylvania survey attracted investment as its basis in scientific facts made capital expenditures appear more certain.[22] Conversely, the Geological Survey of Virginia by Williams Barton Rogers languished, with little interest or state intervention.[23]

In the decade prior to the Civil War, the success of Pennsylvania's coal industry renewed interest in Virginia. Whereas by then Pennsylvania had a robust selection of state-sponsored charters for mining investment, low capital match by the State of Virginia undercut the opportunity for substantial outside investment. New York financiers, for example, had definitive figures in play for judging investment potential and sought state charters close to $1 million. In its legal environment, Virginia also had not developed the kinds of security that Northern or European investors, at that time, sought. Western Virginia had in the interim developed coal mining in the Kanawha and Ohio Valleys, but their output and significance lagged far behind the industry of that time in Pennsylvania.[24]

During the Civil War, the Richmond [coal] Basin was attacked, reducing but not eliminating coal production for the Confederacy. After the war, mismanagement and lack of vision continued to hamper the growth of that Basin.[25]

In 1884, Hotchkiss undertook the republication, under one volume, of William Barton Roger's *Geological Survey of Virginia*. Both Hotchkiss and Imboden were of like mind that had this survey been more deeply considered earlier in their century, Virginia, by their estimation, would have moved more quickly away from an enslavement-based agricultural economy, and, potentially, would have been differently positioned both politically and economically prior to the Civil War. Imboden and Hotchkiss assumed Virginia would have been economically in line with the industrializing North. They considered Barton's departure from Virginia in 1853 for Boston, where he became the first president of the Massachusetts Institute of Technology, a tragedy for science and industry in Virginia. With Barton's departure, Virginia lost its most preeminent scientist and the person with the deepest geological understanding both practically and theoretically of the kinds of mineral-based economy Virginia could develop.[26]

In "Jedediah Hotchkiss, Gilded-Age Propagandist of Industrialism" in *The Virginia Magazine of History and Biography*, Jerry Bruce Thomas depicts Hotchkiss as a fervent evangelist for the commercialization of the natural resources of Virginia and West Virginia to bring forth a new economic era to those states. Like many former Confederate luminaries, after the South lost, due to imposed federal restrictions, Hotchkiss found himself locked out of former investments, particularly in what became the Kanawha Coalfield in the Charleston area of West Virginia, and, he also found himself substantially reduced in social standing.

Though a member of Virginia's Conservative Party,[27] Hotchkiss was anxious for Virginia to leave behind its enslaving identity, for it to shed itself of its antebellum obsession with land, and to turn instead to presumed prosperity in "mines, mills, and furnaces." After trying his hand at many other endeavors, Hotchkiss turned his full attention to reigniting iron and coal potential in the two Virginias. In 1880 Hotchkiss assembled the first journal of what would become four volumes under the title of "The Virginias, A Mining, Industrial and Scientific Journal, devoted to The Development of Virginia and West Virginia" with himself as editor and publisher. In Volume 1, he offers this brief biographical information:

> Jed. Hotchkiss,
>
> Consulting Mining and Civil Engineer; Member of the American Institute of Mining Engineers, and so on—Author of "A Geographical and Political Summary of Virginia," of "A Physiography of Virginia," and so forth. Formerly Top Eng. of "Stonewall" Jackson.

This collected volume, coming in over 200 pages, and published at Staunton, Virginia, combines scientific, technical, and economic information with advertisements for geologists, lawyers, engineers, and also promotes economic advocacy and the solicitation of additional scientific knowledge of the mineral wealth in these two states. Published monthly, Hotchkiss conceived of this journal as essential for rallying the interest of investors.[28]

The four volumes draw from every scientific and technical expertise available at the time both in published papers and descriptive correspondence with experts regarding the two Virginias and mineral exploitation potential. Two columned and dense with graphics, maps, and explanatory text, Hotchkiss, in the introduction, contends that these two states have the mineral resource potential to support growing to be the largest in population among the U.S. states.[29]

Important to this discussion are the aspirations Hotchkiss utilizes science to frame. The Virginias journal's introduction is worth examining at length for its juxtaposition of science, mineral, and other industries, and its postulation of mineral-based prosperity. Hotchkiss writes:

> The development of the resources of the great territory, 67,500 square miles in extent, embraced in the states of Virginia and West Virginia, is the end we have in view . . . there is no region, of equal extent, in the United States, having so much and such a variety of unused natural wealth, and that nothing is wanting but a development and utilization of their resources to enable them to rank with, if not lead, the first of other States in population, industrial activity and accumulated wealth, and to acquire the power derived from the possession of an unlimited store of raw materials with skilled labor and capital to work up and market them. First in importance we place the development of their iron, coal, and a dozen other mineral resources, the extent and richness of which it is even difficult to exaggerate; that done, improvement in all other directions will follow . . . a Mining Journal, striving to collect and publish full and reliable information . . . how they may be made profitable; providing a medium for calling the attention of capital and skill to them, and recording the progress . . . such a journal is greatly needed.

Hotchkiss goes on to discuss other adjacent industries such as timber and agriculture and their prospects in the region . . . and turns again to the focus on science in this journal:

> The basis of substantial material development is scientific knowledge wisely applied, therefore we shall devote a portion of our space to the presentation of scientific facts and statements relating to the Virginias.
>
> We invite communications and correspondence about mineral deposits, the opening and output of mines, the erection and production of furnaces.

It is worth pausing to comment here on Hotchkiss' tying resource wealth to a broad vision of development, industrialism, and prosperity for the two

Virginias, and, to note that the resources he underscores are not only mineral. In the work I do on the ground in this same region with other social change actors, one of the recurring themes of our work has become—mineral resources are not our only source of natural resource wealth here. We have often summarized our search for sustainable economic potential as, *what else would have been developed here if mineral wealth had not come to dominate*? I want to emphasize this as key in thinking about rural spaces, and, about Big Rural more generally. What was orchestrated to dominate, whether coal and now natural gas in rural Appalachia, corn in the rural U.S. Midwest, agricultural production in rural California, pine plantations in rural South Carolina, or, for example, oil and natural gas in the Caspian, has been culturally manufactured to massive scale through scientists and engineers. These productions that now seem essential, matter of fact, have largely been manufactured in the last century to century and a half, and, in total, appear to run so seamlessly that we take their productions, and the people at every level culturally producing these productions, for granted.

Diving deep with Hotchkiss, we see the assertions of science and technology and its labors as beneficial. We look back at Hotchkiss with the knowledge that this hope of his did not play out in the end as he had wished, though, as I illustrate earlier in this volume with reference to the work of Talmage Stanley, for a while, the opening of these coalfields *did* bring people prosperity as long as people were needed to function as technology. Then, mechanization of the processes of mining coal sent the region and its people on a backward trajectory in terms of population growth and standard of living.

Hotchkiss lobbied hard for his case for material, and, state prosperity through mineral wealth exploitation. Among his experts include a range of geologists and professors—prominent among these were John L. Campbell, professor of Geology and Chemistry at Washington and Lee in Lexington, VA; Professor Thomas Egleston of Columbia College, New York, School of Mines; Professor N. S. Shaler of Harvard University, the Director of the Kentucky Geological Survey; and, engineers in the employ of the U.S. Navy.[30]

In *The Virginias* Campbell notes in his section on the "Mineral Resources and Advantages of the Country Adjacent to the James River & Kanawha Canal and the Buchanan & Clifton Forge Railway" that his "explorations have been entirely voluntary" and not for any kind of commercial gain for anyone; they have been accomplished "solely in the interests of science, and the State of Virginia."[31] What proceeds draws from his own research and others', including what he terms the "well-known" work of William Rogers.[32]

Later, he characterizes the compilation of *The Virginias* as especially being in need by West Virginia and that this journal ought to have been created by the state itself and that through the influence of this journal, capital may be raised to develop "the long neglected mineral resources of the two States."

Included in the publication of the *Journal* is a range of testimonies of previewers and reviewers, the *Religious Herald* of Richmond noting that "such journals are invaluable for directing attention to the material resources of the State . . . Capitalists . . . will find in it much useful information." The *Kanawha-Gazette* points out the uniqueness of this journal as being the only one like it that they know of in Virginia.[33]

Hotchkiss reveals significant investment in the Kanawha Coalfield by Swiss investors "The Swiss Commerce Society" and their possession of 35,000 acres there, about 35 miles from "Charleston, the future capital of West Virginia." His enthusiasm for their investment is matched in kind by an enthusiasm for their race and origin. He notes:

> There is no better region in the United States for the location of the hardy and substantial immigrants of Central Europe, chiefly of the Teutonic race . . . the true policy to be pursued in this rich mining region is to sell the surface to a good class of people and raise on the spot a plentiful class of workers—miners, manufacturers, farmers, and graziers . . . such men will be born conservators of the peace . . . but will hear no more of "strikes" and labor riots, the ebullitions of the passion of men that are not free-holders.[34]

Thus, along with anticipating the settling of additional Whites in the coalfield regions and his bias for Northern Europeans (we can assume this was a response to the significant numbers of Catholic Irish and Southern Europeans coming to the United States at this time), Hotchkiss fully anticipates manufacturing, farming, and grazing to continue in the same exact region as deep coal mining, preferably, by Teutonic landholders.

Continuing his promotion of the scientific endeavor of this deep coal mining, Hotchkiss lets us know in the March 1880 issue that many prominent mining engineers had been active recently or were currently active in Virginia and he proceeds to roll call them. Strategically boosting the stature of his own endeavor with "The Virginias," Hotchkiss prints the response he received from William Barton Rogers, the "venerable father of American geology" and Barton's request for a continued subscription of two of each issue, one for a personal copy and one for the "Institute" (The Massachusetts Institute of Technology).[35]

Cataloging every scientist, article, geographic mention, friendly testimony, and investment Hotchkiss curated for *The Virginias* may be generally interesting historically, but in the context of the discussion at hand, these abundant examples in *The Virginias* serve to highlight a range of issues. Just in the samples in the discussion, we encounter the throughline to what I have elaborated with respect to science and engineering and their continued importance in shaping the rural industrial space of the Pocahontas Coalfield today. To recap:

- The symbiotic connection between science and the development of the Southwest Virginia and Southern West Virginia coalfield and other mineral deposits. This rural space remains highly classified, categorized, and a technical space.
- The tie between scientific evidence and capitalist investment in this subsection of the nineteenth-century Southeastern United States. Mine company investors continue to maintain relationships with institutes of higher learning in the region.
- The use of science as a tool to promote capital investment and assure financiers.
- The self-education or apprentice education of scientists or engineers prior to the complete solidifying of, and legitimation of, certain sciences in and by academia.
- The promotion of industrialization as a progression away from enslavement-based agricultural economies and as a solution for the adjacent economic woes of a mostly agrarian economy. A distinctly nineteenth-century concern.
- Engineers also as industrialists, propagandists, and entrepreneurs—in its earliest years, the Pocahontas Coalfield's developers and later coal "barons" often possessed technical or engineering capacities. They not only propagated science, but also shaped community, opinion, politics, and so forth.
- The assumption that farmers, graziers, manufacturers, and miners could and would inhabit the same economic geography.
- The seeking of foreign capital and investment in the West Virginia and Virginia coalfields. This space is an international industrial space obfuscated by tales of wild and wooly back hills men, both in popular depiction and in rural and other studies areas.
- A story of who ought to be in charge, what their optimal ethnic and racial heritage ought to be, and that the right race would keep down the riff raff or unionizers.
- The assumption that mining coal was desirable, preferable, and privileged over whatever other state of affairs and lifeworld of the region's then inhabitants—people, animals, or minerals. There was a purported civic responsibility to this development—and a framing of a time to move from an outdated economic model to a new one more in synch with more prosperous areas of the country.

- An anticipation that the development of the region's mineral deposits would uplift the people of the region, relieving them of their "current" circumstances and that those current circumstances were undesirable.
- That the scientists at the time valued knowledge-production, but many also wore other hats, including that of patriot or economic uplift. Resource exploitation was not only for personal or shareholder gain but for regional or state gain. In Hotchkiss' case, for example, scientific or capitalist duty was not separate from patriotic or civic duty.

A sample plate from *The Virginias*, and other artifacts from the research for *Big Rural*, can be viewed on the accompanying website, www.bigrural.space.

At bigrural.space you will see figure 2.6: In this plate alone, the appearance of, c. 1880, engineers, iron furnace owners, lawyers, businessmen, and investors. This plate reads like a map of many of the Pocahontas Coalfield region's towns' names and those of many of the larger towns' streets, significant boulevards, or landmarks. These people inserted mineral development into the region and inserted their own political positions in this new economy and acquired political cum geographic relevance, their stations forgotten, but their names codified.[36]

The cultural work accomplished by science and technology remains largely obscured, hidden by other kinds of regional and/or even academic, theoretical, or even political narratives. The network needed to create science supporting corporate activity for coal development in the Pocahontas Coalfield remains robust, yet largely unconsidered by scholars outside the industry—and nearly universally unconsidered in social science.

By contrast, in some spaces, the story of science and technology dominates and overpowers other kinds of explanatory cultural forces.[37] We can draw a line from the promises and prides of Hotchkiss to continued questions of rural economic development, especially with respect to the relationship or stances to "progress," prosperity, scientific knowledge, and knowledge ownership, the responsibility of science, and the impact of technology. One tactic toward improved policy and development outcomes in the rural industrial space must include mimicking the scale and breadth of rural industrial intervention but with corporate science not as its base. Instead, the rural space needs actors and ethics committed to people and land for sustained stewardship.

Rational Rural and Patriotic Science

In Hotchkiss and his practices in the emerging science of geology, and his advocacy for a scientific-based industrial intervention in the two Virginias, we encounter a menage of American values (numbered for convenience of further reference rather than implying order of Hotchkiss' or current importance):

1. Hotchkiss the surveyor and geologist rationalizes the land, measuring, quantifying, making nature understandable, and thereby, malleable.
2. Hotchkiss the entrepreneur advocates for measurement as a means also to predict and to anticipate needed and potential capital.
3. Hotchkiss the propagandist utilizes science to sell nature, to rally and to justify investment.
4. Hotchkiss the patriot proselytizes courses of scientific exploration and support as well as courses of industrial and capital investment as acts of devotion to one's country (Virginia). He expects the Pocahontas Coalfield to bring prosperity to the states and peoples of the two Virginias and it is a patriotic calling and duty to lobby for this.

Hotchkiss' faith that the Pocahontas Coalfield would bring economic prosperity to these two sections of the two Virginias I have already referred to as "Hotchkiss' wager": that one anticipates a net positive outcome from a large-scale scientific and/or technical intervention, especially with respect to economic prosperity. Two on-the-nose points worth highlighting include:

- Scientists, engineers, and those people studying them could recognize points 1–3 above as justifications today for scientific practice. Number four may stand as far less of a recognizable reason today for engaging in science, whether for reasons of science being presumed to be above politics or Kuhnian or Mertonian assertions of scientific impartiality and self-regulating community. Moreover, science in the United States currently directly relates to industry imperatives. The many university-corporate partnerships attest to this. In the United States, we rarely turn to models of science for the public good (beyond regulatory science—though climate imperatives may be altering this) or engage citizens in determining scientific research agendas.[38]
- Hotchkiss could not have anticipated that the long-term effects of the Pocahontas Coalfield region itself would result in more net losses than gains for economic prosperity. Given his anticipation of a continued large-scale human workforce and continued surface activities of manufacturing, farming, and grazing, he did not anticipate, nor could he have anticipated, the current state of the state: the extent of surface environmental devastation, the poverty, the health issues, etc.

Let us now review each of these in turn and relate them to current science and technology in the Pocahontas Coalfield.

Rationalizes the Land, Measuring, Quantifying, Making Nature Understandable, and Thereby, Malleable

Maintaining scope with respect to the Western project of rationalizing human intervention upon nature, and American or Western colonial measurement and intervention upon land in North America, can be a difficult task. For example, the states and state borders by which most Americans hold so much stead are human-created interventions, greatly imperfect and impose lines of rationalization of space, territory, laws, geographies, and people. With respect to Hotchkiss, he was both science-making (geology) and state-making (Confederate-surveying and the Pocahontas Coalfield-constructing), and in both cases, science and state-making categorize, rationalize, knowledge-make, and impose means of control over the land, and, its inhabitants.[39]

By way of summary, the Scientific Revolution in the West created a dent in the religious ordering of the world and offered up a world supposedly made comprehensible by science. Simultaneous to this rise of knowledge based on observation rather than divine inspiration an increasing capitalist and middle class arose.[40] As access to scientific research, knowledge sharing, and the instruments to assess natural phenomena were largely sequestered to the upper classes, scientific knowledge and wealth, while not necessarily subdomains of each other, often occurred in tandem. Likewise, in current Western society, the fiscal sponsorship of science requires such wealth that usually taxation provides for scientific and technological development support, the patronage of science characterized as an issue of economic progress, national competitiveness on the world market, and a major concern of national security.[41]

With respect to the rationalization of mining as an acceptable economic practice, in *The Death of Nature*, Carolyn Merchant examines Georg Agricola's *De re metallica*, the first comprehensive defense and technical guide for mining in Europe.[42] Contemporary with Agricola's defense, Merchant explains that in sixteenth and seventeenth-century Europe, the earth "was considered to be alive and sensitive, it could be considered a breach of human ethical behavior to carry out destructive acts against it."[43] Agricola, by contrast, posits mining's relationship to nature as no different in kind from fishing, since both mining and fishing procure what lies beneath the surface for the use of man above. Merchant points out that unlike the material procured through mining, fish renew, though, she correctly explains that Agricola may have had no reason not to believe that the materials mined from beneath the surface of the earth were not also capable of regeneration.

Agricola admonished against a return to a more "savage" time when men were gatherers in the forest, that without mining, men would have to return to this kind of barbarism.[44] Another of his defenses of mining included a response to those people complaining of forests being cut down for mining, that mining is located, usually, *in unwanted places, places useless for anything other than mining* [emphasis is mine—again, and, humans decide this and emphasize one sector over other potential sectors in that space chosen for mining. There are no "unwanted places" or places "useless for anything other than mining."] Finally, Agricola asserts that the greed encouraged by mining does not derive from the act of mining, but from the avarice already inside humans. Mining may bring this avarice to the fore but does not cause it.[45]

Agricola's assumption of mining disturbing what lay below the surface and not disturbing the land above resonates. In *Removing Mountains*: *Extracting Nature and Identity in the Appalachian Coalfield*, Rebecca Scott explains that in the Appalachian coalfield context, the break for many miners in terms of their rationalization of the destructive force of mining altered with the post–World War II escalation of the adaptation to surface mining.[46] Prior, Hotchkiss' contention that farmers, graziers, and miners, and manufacturers could inhabit the same economic geography obtained. Not only did miners also farm or subsistence farm surface land, but they also hunted, accessed family graveyards, and engaged in other outdoor recreation, all while mining occurred below. Local attitudes shifted when surface mining erased historical kinds of surface interactions miners could have with the land. Miners no longer could engage with the surface as they had before—hunting, farming, and so forth.[47]

The technology of surface mining, embodied in its gargantuan machinery associated with and derived from road building, flattened both local land and local, more intimate forms of knowledge. Even if miners had to make the rationalization leap Agricola proposes, and Merchant characterizes as a move toward acceptance of digging around in Mother Nature, deep mining, especially in its earliest years, while dangerous and difficult, also offered a kind of intimacy with the land akin to the kind of intimacy small farmers also had to have with their own acreage, with its microclimates, topography changes, geography, and weather in order to effect the necessary interventions to produce a harvest. Technological efficiencies in deep mining, while improving output, erased or obfuscated many of the intimacies underground mining miners had come to depend on: hearing the changes in the roof, seeing the changes in the seam, communicating to one another about any shifts in any of these kinds of observable markers. Likewise, the huge machines of surface mining took the miner completely out of the ground, placing him (it was usually a him) in a machine far above the ground the machine was hewing, and its technical capacity required little to no foreknowledge of space: any mountaintop could be flattened in a nearly similar manner until its mineral bounty was exposed. Geologists provided a map of how far and how deep. One did not need to rely upon developed knowledge perfecting what scientists aka geologists predicted, as in the case of earlier deep mining.[48]

Flattening the Pocahontas Coalfield

If science creates knowledge, then these large machines, in the case of surface mining, destroy a kind of knowledge, an intimacy of and practice of place that scholar James C. Scott likens to the Greek concept of "metis"—a practical, intimate knowledge.[49]

To broaden the perspective on this usurping of intimate knowledge by large technical machines or large-scale interventions—deep mining indicated a layering in of additional possibilities of knowledge, coexisting with then-current knowledge and practices of farming, grazing, and manufacturing—all activities that continued at the time in the Pennsylvania coalfields.[50] Though STS theorists propose that transactions of knowledge now take precedence over natural resource wealth, two facts remain current—natural resources continue to matter, especially energy resources, and science and technology were key in their exploitation during the nineteenth and twentieth centuries. While scholar Sheila Jasanoff, for example, engages the politically new territory of managing and politically maneuvering among governments and forming a State relationship to biotechnology, I contend that studies of the sociology, history, philosophy of science and technology have focused too much concern for too long on certain elite sciences and technologies: physics, genetics, computer technology, nuclear technology, biotechnology.[51] Given what I have outlined with respect to science's enabling of the Pocahontas Coalfield and the symbiotic relationship of corporations and science in this rural industrial space since the space's inception, blanket statements about science and its newly emerging close relationship with power and politics leave one scratching one's head at analysis such as this:

> Science and technology have been regarded for centuries as instruments of social progress and personal liberation. Yet, as scientific knowledge becomes more closely aligned with economic and political power, producing new expert elites, the distance between the governors and the governed can be expected to grow—a dismal prospect in societies where low levels of electoral participation and citizen engagement are already causes of concern. Science, moreover, has historically maintained its legitimacy by cultivating a careful distance from politics. As state-science relations become more openly instrumental, we can reasonably wonder whether science will lose its ability to serve either state or society as a source of impartial critical authority.

While Jasanoff certainly is right that biotechnology opens new lines of concern with respect to access and equity, in the Pocahontas Coalfield, long-standing relationships between governments, corporations, and science already account for the general disparity she outlines. One cannot imagine that she naively assumes only science that has adhered to the idealized form that she also describes remains the only legitimate "science." Yes, as she examines, it is correct to investigate whether biotechnology will create a new "genetic underclass"; yet I contend that other sciences such as geology and mining engineering already have contributed to the creation of underclasses and continue to do so. In the case of the Pocahontas Coalfield, the culture of scientific objectivity turns a blind eye to the material-metis erased not only in the form of replacing people with large machines but also to the erasure of human capital and intimate knowledge, a willful ignorance of or accountability for "impartial" scientific knowledge's economic and environmental consequences. Taxpayer funded and state allocated support for this unfortunate dance of science and technology has long been a part of this practice.

I concur with Jasanoff that the discussion of democratic society must include an examination of the politics of science and technology. Where I differ from Jasanoff is that this imperative is a new phenomenon with respect to science.[52] Whether going back to the necessary support of the Virginia State Legislature for William B. Roger's survey work in the 1830s or to the intermingling of democracy, research, invention, and state-making coproducing during the time of the creation of the United States, an examination of science and technology and state-science relations may not always have been part of what was examined in reflections upon post–Enlightenment democratic theory, but it ought to have been.

The natural-resource-based society remains central with largely rurally produced resources affording the metropole the luxury of focusing on and managing knowledge production. Given the continued population growth of the world, and the destruction of the environment to sustain growth economies to support this population expansion, we are entering a new knowledge society that potentially renders this wealth transfer from rural or offshore to metropole as continuing to be black-boxed, invisible, and thereby making the raw material resources and the people involved below the class level of knowledge worker also invisible, disenfranchised. The invisibility of Big Rural abounds.

Any number of academic fields and literatures contend with this question of "ought we?" in science or political policy. For example, because we possess the technological capacity of massive nuclear weapons, ought we produce them? Though a decision tree indicates more economic efficiency in utilizing surface mining equipment, ought we? The risks associated with this large-scale technological intervention outweigh the benefits, or, at least are unknown in the long-term, thus, ought we follow with precaution? Would we engage in this large-scale technical intervention if it were in a city center, in the first world, in full view?

The introduction of technology can be layered in meaning, and the advent of technology can vary in principle meaning from era to era. For example, in agriculture, simplicity of processes was supposed to be good and complexity was supposed to lead to failure. However, as James C. Scott demonstrates in *Seeing Like a State*, Western large-scale agricultural intervention flattened local knowledge and flopped in its production capacity. In these cases, the narrative we tell ourselves about science and technology can also shift—we discover the evidence that reveals complexity as desirable, that certain scales of objects dehumanize us and result in unanticipated consequences. We find ourselves having to attempt to recreate and mimic unplanned prescientific forests to replace the complexity lost through scientific and technical interventions.[53]

Hotchkiss' wager, a massive industrial intervention, has had long-term local consequences in human and environmental costs that have turned out far from the prosperity he proposed. In this respect, for large-scale industrial interventions, the Pocahontas Coalfield stands not as an exception, but as the norm. The handwringing and puzzlement as to the long-standing economic, social, and health problems there rest with assessing this region in a fishbowl rather than likening it to other failures of scale like those with which James C. Scott contends such as Soviet collectivization, compulsory villagization in Tanzania, or the city-planning of Le Corbusier. Scott just as well could have turned his critique of social engineering and rural settlement and production to the Pocahontas Coalfield. The Pocahontas Coalfield was economically productive for shareholders, yet as a technical intervention that created an industrial coalfield region out of a place where this did not exist, it failed as a local driver of long-term prosperity for that space as a community. People were used as technologies until they no longer were most desired or efficient and thus replaced by machines. As an environmental intervention, given deep mining's coal processing and the railroad's accompanying chemicals, and the scale and erasing of mountains and metis through surface mining, this space also failed. On many accounts and many fronts unforeseeable to Hotchkiss, the Pocahontas Coalfield proves Scott's directive that "any large social process or event will inevitably be far more complex than the schemata we can devise, prospectively or retrospectively."[54]

Why does it matter if the Pocahontas Coalfield is not associated with scientific, technical, or technological intervention in the broader U.S. culture, possibly, beyond anti-surface mining activism? Here are only a couple of reasons why this matters.

Black-Boxing the Pocahontas Coalfield

In the U.S. context, spaces popularly associated with technology or science can be considered more sophisticated (such as Silicon Valley), worthy of time and investment, worthy of discussion—a worthy place.[55] Though the largest machines on earth are employed in rural spaces, the most "worthy" machines in popular discussion are in metropoles ("high tech" such as the software technology industry, or artificial intelligence). In STS we may investigate whether technology has a politics vis-à-vis Langdon Winner but, in the context of party politics—kinds of technologies certainly do. That Silicon Valley had the ear of the Barack Obama administration and that the energy industry subsequently had the ear of the Donald Trump administration registers as news to no one.[56] More complicated is the lament by the latter industry that it (the energy industry) has not done a "good job" of telling its story.[57] More likely, it is the former that is more adept at covering its trails of tears.[58] Part of what makes that Silicon Valley technology compelling is that although it is not as ubiquitous as energy technology, its scale seems much more human: though coal power or gas power may fuel my handheld device, those technologies come to me as public utilities or as ubiquitous goods; I do not notice when I use them; I only notice the energy's absence. That this energy exists has become black-boxed in current U.S. culture.[59]

The people and the regions where this energy originates also become black-boxed, invisible, and taken for granted, and when these regions shed people as technologies, they appear on the national radar only if politically useful (one thread that unites these people in the spaces that constitute Big Rural). In the juxtaposition of the offshore servicing the metropole, not only the Pocahontas Coalfield specifically or even conspiratorially becomes rendered invisible, but also the large-scale industrial interventions in rural spaces across the country become invisible: the biggest machines on earth produce coal for energy for powering my handheld device; large-scale combines produce food for making the sandwich I order from a machine at a convenience store, and so on. The Pocahontas Coalfield and the many other rural spaces like it support the larger project of the role of rural spaces in the United States, functioning as a kind of "large technical infrastructure and support system" (LTS).

For example, in "Large Technical Systems and the Discourse of Complexity," sociologist Bernward Joerges indicates that governments in the West create dependency upon the growth of their LTS to assert financial viability.[60] Though Joerges insists on the material and machine operations of LTS, he admits that splitting hairs in terms of what constitutes LTS in his sense could be applied willy-nilly, as aspects of various machines or systems may overlap or fit into several categories. Nevertheless, his definition of what makes a system large obtains if applied to rural spaces as industrial material-good producing spaces outside the metropole—including a rural industrial space like that of the Pocahontas Coalfield.

According to Joerges, the large LTS means: the technical system is complex with "heterogeneous systems of physical structures and complex machineries." Surface mining equipment, literally hidden from view from most Americans, links to an energy system that is black-boxed, and so ubiquitous as to be out of view in plain sight—stringing us all together by power lines, transfer stations, electrical outlets we assume as landscape, so prevalent we can ignore them.

These physical structures and complex machineries contain materiality, intersect, and intertwine "over large spans of space and time, quite irrespective of their particular cultural, political, economic and corporate make-up." My note: The energy produced from coal does this—it and other energy often appear from the rural into homes or spaces of any sort as if by magic: only noteworthy when absent.

LTS supports or enables many other LTS: energy systems enable communication systems, transportation systems, scientific systems, and so on.

The larger the technical, machine, and physical capacity of the LTS, the larger the political negotiation and decision-making, with much of the politics also played out in science communication and technical terms. "Smaller-scale, local, intraorganizational technical systems" and may evolve from or integrate with other LTS; LTS include "many organizations." "Those dominant actors in LTS who own, regulate or manage parts of them will be coupled . . . politically, financially, and legally." My note: Is coal a national security or energy security issue? If so, how? Is it saving our nation from supporting terrorists? Is it cheap energy fueling our economy? Is it energy that is expensive in the long-term and killing our planet? Is it a miracle in its abundance, a murderer in its use? A blessing? A curse?[61]

LTS "seem to surpass the capacity for reflexive action of actors responsible for operating, regulating, managing, and redesigning them" yet these systems still maintain stability. My note: This plays into a rhetoric of "we are just doing our jobs." This is the job there is, so I do this job. Someone must do mining engineering, reclamation science, energy engineering, geology—these are, after all, jobs.

LTS, as long as they work, are assumed to function and are not objects of much larger public consideration and usually only then on how they have failed. LTS can be "silenced" and "hidden away" and only in their failures, after they have been implanted in society, are they rendered visible. My note: The public scrutinizes failures if investigative journalism reports them; otherwise, out of sight, out of mind. The Pocahontas Coalfield fascinates through its human failures, its poverty porn stories, and occasionally, through its environmental flops. Otherwise, the Pocahontas Coalfield cum rural industrial LTS remains invisible across the country and to the country—one more LTS strung together in the ubiquity of Big Rural.

Issues with LTS often become externalized. Even if a social or other problem may be related to the LTS, these problems can be pushed outside the boundaries of the technical workings of the LTS. My note: The problem, of course, is not metropole energy usage, or food demands, or lack of metis or re-skilling for one's own energy or food needs, or grid inefficiencies but "evil" energy companies, "ill-informed" coal miners, Donald Trump, Monsanto, or evil, "lazy" city dwellers or unrealistic expectations—how else will you feed the world or power the planet without large-scale interventions like genetically modified organisms (GMOs) or nuclear power or oil tankers, or so cry their advocates.

Though aspects of LTS can be dismissed or disappear, LTS cannot exist without considerable social and societal upheaval. My note: We can end rural LTS only through significant societal, cultural, and political shifts. They are not impossible, only improbable unless LTS fails domestically in concert with access to LTS resources abroad.

Reflecting Again on Big Rural

Here I highlight again two significant points: the Pocahontas Coalfield captures national interest for its human or other failings. As a functioning part of the LTS of energy systems in our country, systems large scale and ubiquitous, it and its story of science and technology and people as technology remain rendered invisible. I term this rural industrial system "Big Rural" to ally it with what has become common parlance regarding other large networks or systems, which have become "too big to fail" such as big Ag, big science, big government, big pharma, or even big city/ies (far more powerful than their opposite in kind and thought—small towns). This is a shift from and in contrast to, recent popular and folk conceptualizations of the rural in hyperlocal terms, a "think global, act local" bumper sticker politics, devoid of a deeper consideration of the ethics of either thinking globally or demarcating ethically "local."[62]

Advocates for measurement as a means to also predict and anticipate needed and potential capital. And: propagandist utilizes science to sell nature, to rally and justify investment.[63] These two analyses of Hotchkiss' scientific and surveying practices and science, technical, and investment advocacy resonate with the aims of corporate science today. In STS we debate what counts as science, and what the norms for science are, or have become. As scientific research generally requires significant capital, that capital can come to scientists in several ways:

1. On payroll at a corporation
2. On payroll at a university
3. On payroll with the state
4. Self-funded or funded through other endeavors
5. Patents and so forth generation

With respect to numbers 1–3, tax incentives, research agendas, corporate advisorships, and so on can cause the research agendas of these three to intertwine and conflate. We already have these exhibits in the *Preface* with respect to the Pocahontas Coalfield. Obama had his Silicon Valley advisors; Trump had his energy sector advisors: science shifts in relation to what politics enables. The science remains constant, yet the U.S. signs, then drops the Paris Climate Accords, then signs them again.

The mechanism of professional standards in science obtains in the academic research realms. Scientists adhere to scientific ideals: "modesty, simplicity, straightforwardness, objectiveness, industry, honesty, human sympathy, altruism, reverence"—like perfect Calvinist piety. Yet, science is science when practiced in public, for professional scrutiny, along with professional recognition and promotion. Science shifted from science for science's sake or knowledge's sake . . . to government giving for science without it interfering in science.[64] This rings no less true in mining engineering than in other scientific fields—it also consists of awards, accolades, friends helping friends, and corporations advocating for research funds for certain sciences over others.[65] Nevertheless, what counts as science worth advocating for matters: Hotchkiss may have wagered on the wrong intervention, but his road to hell was well-paved.

Scientists Taking Responsibility in the Rural Industrial Space

Scientists must also consider the environmental, economic, and human costs of their knowledge creation. To fully address how would require an additional book; however, to bring this down to earth, let me briefly consider this in the Pocahontas Coalfield context.

Reclamation scientists could engage local citizens in learning to monitor their own home sites or the streams, land, and public spaces they use in order to understand fully the issues of long-term moving contamination in the rural industrial space.

Geologists could work with local citizens on understanding the hazards and benefits of remaining geological formations and mineral resources and to explain what may be of continued corporate interest or state interest.

Biological systems engineers could teach citizens how and when to monitor their ground and surface water to understand the issues of this rural brownfield space and water's shift from weather and season and the potential for contamination.

Economists could work from models of smart shrinkage rather than models of endless growth. Likewise, they could assist with economic modeling or work that cannot be automated away.

That one can create a great technical intervention may be great "pure science" but that one should or should not is how science ought to be considered in society, in the community, and in the polity.[66]

In the case of Hotchkiss, we have science in service of society. His caveat, however, is science for patriotic ends. Even if all his ends were not for the good of the state in their intentions, the question of science and its service remains valid: what ought science serve?

In a democracy, all state policy ought to consider first whether an intervention suggested supports or brings about more democracy. Democracy and its deficit can be measured. The following brief case study brings into close relief a space adjacent to the Pocahontas Coalfield that illustrates this moral-scientific and democratic policy issue.

Corporate Science and Technology and Democracy Deficit in Lindytown, West Virginia

Named for Charles Lindbergh, the unincorporated town of Lindytown, West Virginia sits in the Kanawha Coal River Coalfield, north of the Pocahontas Coalfield. At minimal issue in the case of Lindytown is the World Bank-defined basic obligations of a local government to "ensure the delivery of quality social and communal services and a responsive supply of land and housing, so that the city becomes and remains a hospitable place for businesses and households." The local government, in this case, the county or state government, has not sufficiently protected Lindytown.[67]

Boone County, home to Lindytown, as recently as 2009, had the single largest concentration of mine-related workers in West Virginia and received significant tax revenue from the mining industry. That same year, a subsidiary company of the coal company Massey Energy (Massey has since been purchased by Alpha Natural Resources, a donor to Virginia Tech, among other places) procured from most remaining citizen-residents of Lindytown their property in exchange for these residents' agreeing to leave, and, also, to submitting to not taking Massey to court, testifying against Massey, requesting for Massey's investigation, or speaking against Massey regarding local Massey mines.[68]

According to the April 12, 2011, *New York Times* article, "As the Mountaintops Fall, a Coal Town Vanishes," Massey Energy was motivated to buy out citizens of Lindytown due to the risk of their living in close proximity to Massey surface mining sites.[69] Rather than a central place of commerce or multilayered resource-tied economic endeavors, again, in contrast to what Hotchkiss wagered—this section of West Virginia became hidden, and elsewhere, like the "throwaway" and hidden places of mining in Africa (far from the users of the mined materials) or historically in South America (also, historically far removed from the main recent users of the minerals mined there).[70] In order to extend the theoretical conceptualization of the effects of large machinery and large technical systems erasing local metis, I turn to a postcolonial interpretation of the destruction of people and place, or removal of people and life, in order to exercise corporate (state) will.

Lindytown as Read through Achille Mbembe's "Necropolitics"

According to Achille Mbembe, necropolitics and necropower concern the contemporary manifestations of "resistance, sacrifice, and terror" which lead to "deathworlds" in which people are oppressed through exposure to the ultimate in weaponry and control through weapons, leading to an existence that constitutes not much more than being dead while alive.[71] Furthermore, the power to decide life or death no longer rests with the state but becomes scattered, residing in other formal or informal social or institutional forces. For Mbembe, scattered necropower blurs the boundaries between uprising, self-murder, recovery, suicidal martyrdom, and liberty.[72] In particular, he traces the interplay of these concepts as constituting necropolitics and necropower in Africa, supplanting earlier forms of colonialism concerned with terror as discipline and control, though this earlier colonialism set the stage for the current extremes of necropower.[73] For Mbembe, this control of life through death or death-like life extends Michel Foucault's concept of biopower, or, the extension of the state's sovereignty concerning who must live and who must die.[74]

In this examination through Mbembe's lenses of necropower and necropolitics, I evaluate certain of his concepts as potentially applicable to theorizing Lindytown. Before turning to this explication, we must first become better acquainted with the situation of Lindytown.

Lindytown by Barry, Biggers, and Letson

The two newspaper articles I employ for this examination create a composite of Lindytown's remaining residents as terror-stricken, and the town as having been purposely dismantled by Massey Energy's mining practices well before the remaining residents sold their homes. Through descriptions and quotations, the journalists create a sense of relentless anxiety. *New York Times*' writer Dan Barry sets the stage of a place lost, "here just a moment ago," created and destroyed by coal to support "our" way of life, by which I am uncertain if he includes the first-person plural to also indicate the way of life led by the remaining or former inhabitants of Lindytown or he only means those people (us) benefitting from a Lindytown as a type. As his illustration of the people that remain in Lindytown, Barry focuses on the Richmonds, an elderly couple who have opted to stay as has their retired coal miner son. Behind the Richmonds' home, mine-blasting has exposed a huge rock. Barry notes that the elderly Mrs. Richmond has Alzheimer's and returns to the back window throughout the day to check on the rock, lest it finally teeter loose to destroy their home and them.

Barry recounts the explanation Massey provides for Lindytown's demise, that residents wanted to move, and thus asked Massey to purchase their homes. According to Massey's general counsel, Massey's mountaintop mining in Lindytown was well within legal limits, and Massey bought these properties only as "additional back up to state and federal regulations." Barry counters this perspective with the words of local retired miner James Smith that people wanted to leave Lindytown because "the mountaintop operations above had ruined the quality of life below." The miner posits that it was most likely less expensive for Massey to purchase the community than to address these quality-of-life issues or face legal action over silica dust and other contaminants. Smith continues that he would have stayed had the mountaintop removal not made Lindytown uninhabitable, commenting that, "You might as well take the money and get rid of your torment. After they destroyed our place, they done us a favor and bought it."

According to Roger Richmond, the Richmonds' son, people left because they "were tired of fighting" or "of having to put up with all the dust. Plus, you couldn't get out into the hills the way you used to." For example, due to government regulations of surface mining, should Roger Richmond seek to visit a family cemetery now located within the mining zone, he would "have to make an appointment with a coal company, be certified in worksite safety, don a construction helmet and be escorted by a coal-company representative."

In Jeff Biggers' "The Coalfield Uprising" in *The Nation*, we meet Steve and Lora Webb, who leave their home directly outside Lindytown on the eve of the Environmental Protection Agency's (EPA) declaration on September 11, 2009, of the violation by all "pending mountaintop removal mining permits in four Appalachian states" of the Clean Water Act. Massey installed a twenty-story dragline[75] near their home and as part of this mining operation, blasted rock twice a day, which sent coal dust and fly rock[76] into the air, covering the Webb's house and property. Biggers quotes L. Webb's words,

"It's unreal. It's like we are living in a war zone." By the spring of 2009, the Webbs were one of the last families in or around Lindytown, and Biggers reports that the Webbs appealed to state and federal agencies for the enforcement of mining laws. According to the article, neither level of governmental agency responded with action, prompting the Webbs to finally sell to Massey after blasting near their home was temporarily halted by federal regulators, then allowed to resume, even closer.

Biggers cites S. Webb's concerns that though the federal government under Barack Obama seemed more responsive than the Bush administration to concerns such as his own and the impacts of mountaintop mining, he worried that Obama made compromises like those made by Jimmy Carter, which led to the expansion of surface mining in 1977. Biggers then extrapolates what these compromises may be, such as the EPA's issuing of new permits under the guise of economic benefits. Further, Biggers cites research that some streams contaminated with mountaintop slurry have decimated the local population of mayflies, which is equivocal with the uninhabitability of this area by humans. According to West Virginia Department of Environmental Protection biologist Doug Wood, "The loss of an order of insects from a stream is taxonomically equivalent to the loss of all primates (including humans) from a given area. The loss of two orders of insects is taxonomically equivalent to killing all primates and all rodents through toxic chemicals." Biggers moves from there to a description by activist Chuck Nelson of an action planned by the Ohio Valley Environmental Coalition to hold a vigil for Lindytown, quoting Nelson, "I guess you could call it [the planned action] a funeral, for all the families that used to love this land and considered it home."

In Al Letson's episode "Appalachia Rising" for the *State of the Re*: *Union* radio- and web-cast, the principal respondent includes Boone County resident and environmental activist Maria Gunnoe.[77] She explains why many residents leave as pertaining to having no protection with respect to air quality or any means to safeguard their health from the coal dust from surface mining. As Letson and Gunnoe enter Lindytown, Letson describes Lindytown as reminding him of what he had seen of New Orleans's Ninth Ward after Hurricane Katrina. The houses have boards over the windows. There are insect sounds rather than the sounds of a town. Many of the doors to buildings and houses stand wide open, some homes still displaying remnants of the families that had lived there. The main difference, though, according to Letson, is that, unlike the Ninth Ward, the homes in Lindytown are still habitable.

Gunnoe explains that Route 26 to Lindytown used to be "a through-road, now it dead-ends in a mountaintop removal site. The coal companies have bought up all the communities that used to be up here. They have got it down now to where there are only three communities left on Route 26." She recounts the last days of Lindytown:

> November of 2009 the people of Lindytown were given sixty days to move out. The coal company bought up all of their houses. Initially they were told they would be able to stay in their houses for an undetermined amount of time. Then, once they [Massey's subsidiary] became owners of this real estate here, they gave the folks sixty days to get out right in the middle of the holidays . . . the purpose of buying that town is they are blowing that mountain up which is the backdrop of that town. They can't have people living at the bottom of this mountain because then all of a sudden it goes from mining coal to murder.

Again, in this reportage, we meet the Richmonds, the town's remaining residents. Recorded before the elder Mr. Richmond's passing, we hear his own words with respect to his choice to remain in Lindytown:

> We hear blasting here about every day you know. Sometimes it's loud and you feel it right in the bottom of the floor to your feet. . . . You wouldn't think that an individual or individuals call this home, in a place as desolate as this at this time. This neighborhood here has become part of me since I was just a young fellow, since World War II, you know. When they wanted me to move, I decided that I didn't want to move. Sentimentally this is just as important to me as what that coal is up on the hill for Massey Energy, you know.

Gunnoe punctuates Mr. Richmond's narrative by commenting that although Mr. Richmond served in World War II, he "doesn't even have the rights to protect his own home." As Letson and Gunnoe visit the Richmonds, we again get commentary regarding the boulder perched above their home and its possibility of falling at any moment. Letson also paraphrases one of the respondents that West Virginia is a national sacrifice zone, where residents are forced to live under hard circumstances so that the rest of the nation can benefit.

Roughly three miles from Lindytown sits Sundial, West Virginia—home to Marsh Fork Elementary. Here we meet Ed Wiley, who worked a large portion of his life on heavy equipment on surface mining sites, and during this time he had been pro-surface mining. His perspective changed, however, when his granddaughter became sick, and he noticed many more children from her school were also out sick. Despite working in the industry for years, he explained that he learned facts about coal that he had never known, that is, that coal contains deposits of lead, arsenic, cadmium, and uranium. His granddaughter's elementary school sits 250 feet from a coal sludge deposit—a place for the runoff from coal once it has been cleaned to be shipped for industrial employ. Concerned for his granddaughter's health as well as that of the other children, he contacted the county school board, the health department, and other government agencies, prompting him to claim, "Everybody I turned to that was supposed to be responsible or had the power to help our children just wasn't going to do nothing. Nobody wanted to touch this issue because it was coal."

Only through activism and outside attention were Wiley and the people he organized able to push for relocating the school. Local politicians did not move unless forced or shamed by power originating outside the community.

EXTENSIONS: NECROPOLITICS AND LINDYTOWN

Extension 1: Sovereignty

Given the descriptions by the handful of remaining and former residents of Lindytown of the choice to leave as not an authentic choice but one forced upon them through the risk to their health and the potential for an untimely demise by staying, the word "choose" becomes synonymous with "forced to." Again, one of the key concepts in a liberal democracy rests with the idea of being a free agent, not bound to a particular person or bound to the land: one is free to choose where to live, and, to live.

Further, in the current context of Lindytown, the word "citizen" loses its power to denote an active, free agent when paired with "of Lindytown." The individual inhabitants of Lindytown did not and cannot make active, free choices with respect to the fate of their town. In the case of deciding the fate of Lindytown, through Massey's making remaining synonymous with choosing an early death, the individual inhabitants of Lindytown are stripped of their citizenship; should they stay, their lives would become lives-in-death; by leaving, they ensure the death of Lindytown.[78]

Prior to World War II, in the European juridical order that predated necropower, authority topped out at the level of the nation-state; no one body acted as a sovereign above that of the level of the nation-state. At the level of nation-state, the state possessed the right to wage war on another nation-state, and, within its own borders, take life. In the case of Lindytown, the sovereignty leaves the level of the nation-state and has moved to another agent, Massey Energy.

Extension 2: State of Exception

The death of Lindytown qualifies for Hannah Arendt's explanation of totalitarianism. To be a citizen of Lindytown is an oxymoron. Lindytown is no longer a "town" and no longer has citizens; it has entered into a "state of exception," a relational position at the extreme of our rational conception.

The environmental degradation accompanying surface mining stretches to the extreme of what we allow to occur, yet, the place where it occurs is not just anywhere; it is a place devalued on several fronts: peripheral, rural, ubiquitous in its kind of technological production and would-be associations (energy—energy is everywhere—it just appears). Former inhabitants and inhabitants speak of surface mining's effects on Lindytown as being like a state of war.

Extension 3: Politics as the Work of Death

Civil discourse in Lindytown is nonexistent or suspended. Politics in Lindytown is not dialectical. There is no creation of common understanding through shared discussion toward the rational. Shelia Jasanoff's civil and civic science turned its back on Lindytown; the taboos against violence have been broken with the killing of Lindytown.

Manufactured states of emergency contributed to Lindytown's death such as fear of losing jobs (they left in bulk with the advent of automation of the coal industry in the 1950s).

If you have no people, you have no one you need to employ. Eliminating Lindytown strengthened Massey Energy's fiscal security.

Extension 4: The Enslavement

The choices left to inhabitants of Lindytown have rendered them slave-like. They have suffered the "triple loss" of the slave: home, loss of rights over one's body, and loss of political status.

Being a member of a community implies the right to exercise speech and thought. If you face death by remaining in a community, your ability to speak for that community also dies once you leave.

To be enslaved is to be left alive but in a state of injury. The Richmonds live with the threat of death from coal dust or the boulder shaking loose during a blast and killing them.

Mbembe quotes Arendt, that during the two world wars, Europeans treated each other as they had been treating the colonial "savages."[79] People in Appalachia have been characterized as savages, in deficit.[80] Despite my previous interrogation of the term "indigenous" being used by scholar Talmage Stanley to describe White inhabitants in the region prior to the construction of the industrial coalfields in Appalachia, it could be in this common point of destruction of the environment, and denigration of "the barbarian" as justified, that the stories of White *and* Black coalfield inhabitants and colonial indigenous peoples meet, if they do meet.

In the narrative of the logic of heroism, a hero defeats an enemy and thus attains glory. In the logic of martyrdom, the future collapses into the present in a desire for eternity; the body becomes a weapon; the material dissolves into eternal life.[81]

However, in both the logic of heroism and the logic of martyrdom, the death of an enemy is achieved.

In the logic of survival, and in the case of Lindytown, little freedom to live remains if you do not take the buy-out. You have left the freedom to live in the shadow of impending death. You do not even have the choice of a martyr, as in your death, you will not take down the perpetrator with you. Even a slave's suicide hurts the owner economically by denying the master labor. In Lindytown, choosing to remain may be meaningful to a resident—but it does no damage to Massey Energy.[82]

Mbembe quotes Martin Heidegger: "One is free to live one's own life only in so much as one is free to die one's own death."[83] This can be applied to towns. Lindytown, however, stands as no American exception in the rural industrial space.[84]

Extension 5: Science as Knowledge Creation

In his elucidation of necropower and necropolitics, Mbembe issues a critique of scientific logic and reasoning, what he terms the "generalized instrumentalization of human existence and the material destruction of human bodies and populations." For this critique, he utilizes Foucault's supposition that the roots of the state of exception in Europe (World Wars I and II) ran deep in terms of ethnic bias and nationalism.[85] In both World Wars I and II, the creation of technology enabled the capacity to exercise the state of exception (e.g., efficient gas chambers, transport, etc.). With respect to Lindytown, the massive equipment and the explosives employed reduce the number of people mining behind Lindytown but vastly increase the power to take down a mountain and engage in coal extraction. However, the "generalized instrumentalization of human existence and the material destruction of human bodies and populations" and Foucault's "ethnic bias, nationalism, and so on" are not exactly the same things, though I can follow the logic that if engineers and scientists create technology that can be used to exercise a state of exception based on ethnic bias, nationalism, and so on, though they themselves may not actually push the buttons or work the levers, they become complicit in the state of exception.

Political theorists and philosophers such as Karl Popper or later, Yaron Ezrahi, may counter that the state of exception becomes enabled because we have not been too logical, but rather, not logical enough.[86,87] Yet, I am aware that the construction and the conceptions of what counted as "science" in examining World War II as a state of exception may not count as science today (think Eugenics). In this respect, the science used to support surface mining may not be more political than scientific (rational, objective, peer-reviewed, created in a scientific community). Mining requires input from geologists and needs scientific expertise to meet regulatory demands. These are facts, not political viewpoints. Ezrahi also might point out that, now, since countering science has arisen to react to the science of surface mining, the issue has become less political and more reliant on fact.[88] Science in reaction to coal mining enables a more potent strategy based on reason and instrumentalization to arise to counter a more political platform. Further, according to Ezrahi, "once the resources for depersonalizing and objectifying public actions are depleted, actors find it more difficult to persuade their audience that their actions do not stem from personal or partisan political considerations."[89]

Political theorist Ulrich Beck, in *Risk Society*, points out that the democratic apparatus of instrumentalized correction upon which Ezrahi relies to function to counter bad science or bad politics falls short when faced with global risks such as acid rain.[90] Certainly, as unveiled above in the *Extensions*, democracy already has failed the people of Lindytown, and very potentially similar rural industrial spaces not even under such extreme physical duress regarding their people or environment.

Even if science as posited by Ezrahi can be counted on to check morality as well as arbitrary and political actions, the science, or policy, countering the force of surface mining has come too late for this community in Boone County, West Virginia.[91] This totalitarianism also persists in various ways, also remaining invisible to the metropole United States.

In Europe and the other states of exception with which Hannah Arendt concerns herself, only another massive intervention swayed the course—James C. Scott reveals the destruction in this course of action: my examination and focus on the Pocahontas Coalfield, likewise. Though Lindytown has not derailed the larger project of American democracy, democracy there has been derailed.[92]

This brings us back to science and technology policy—or policy in a democracy; any kind of state (communist, totalitarian) can have a science and technology policy. The main questions become, how is that policy democratic policy and how is it enacted to further enable more democracy and more liberty? And, as in the case of Hotchkiss and his wager for large-scale intervention toward creating more prosperity, what then in the realization of policy serves the people, long-term?

In the U.S., we have an active federal history of making explicit the purpose of certain of our policies, and those policies subscribing to our actions. For example, what if the policies the United States purports to support abroad, in its United States Agency for International Development (USAID) mission, were also what the United States supported domestically, or promoted and supported science and technology research to investigate, innovate, and advocate? These would be a strong set of democratic strategies to also promote domestically. *See appendix E.*

CONCLUSION

In *chapter* 2, I assert the implications for science and technology as practices in a single-sector rural industrial space, where they contribute to the denigration of the citizen-resident. This citizen-resident experiences the fruit of scientists' or engineers' "knowledge creation" as political, economic, social, and environmental derogation. Work, and thereby, community, is displaced. Large machines flatten the land and the local understanding of that land. As parts of a large technical system, local problems in the rural industrial space become invisible to society's metropoles, contributing to rural industrial spaces as places both ubiquitous and obscured, places constituting Big Rural, with citizen-residents with diminished civic power, but with resource extractive and production phenomena too necessary to (be) let fail. In the single-sector rural industrial space, democracy suffers a deficit through the destruction of a habitable environment and through the conflation of work and civic life.

By contrast, policy examples exist that would set a more equitable course for government and a range of other policies for citizen-residents in places like the Pocahontas Coalfield or Lindytown. Education, research, science, technology, and so on could follow a policy highlighting first the call to democracy and the creation of a "free society." Science can be no less scientific, no less technical if it also responds to civic duty.

NOTES

1. Jerome R. Ravetz, *Scientific Knowledge and Its Social Problems* (Oxford: Clarendon Press, 1971).

2. Rebecca R. Scott, *Removing Mountains: Extracting Nature and Identity in the Appalachian Coalfields*, Quadrant (Minneapolis: University of Minnesota Press, 2010). Jerry Bruce Thomas, *An Appalachian Reawakening: West Virginia and the Perils of the New Machine Age*, 1945–1972. Vol 12 (Morgantown: West Virginia University Press, 2010).

3. Other scholars cite later dates and evidence for the influx of interest and capital but the primary sources I use point to Hotchkiss' pre-dating those.

4. Anthony F. C. Wallace, *Saint Clair: A Nineteenth Century Coal Town's Experience with a Disaster-Prone Industry* (New York: Random House, 1987).

5. Ibid.

6. Edwin Emanuel Holm Jr., *Production and Marketing of Coal in Virginia and the Nation* (Richmond, VA: Virginia Division of Industrial Development and Planning, 1955).

7. Jerry Bruce Thomas. "Jedediah Hotchkiss, Gilded-Age Propagandist of Industrialism." *The Virginia Magazine of History and Biography* 84, no. 2 (April 1976): 189–202. The author later wrote *Reawakening Appalachia.*

8. Aaron M. Sakolski, *The Great American Land Bubble: The Amazing Story of Land-Grabbing, Speculations, and Booms from Colonial Days to Present Time* (New York: Harpers & Brothers, 1966).

9. Ibid.

10. Michael Fitzgerald, *Splendid Failure: Postwar Reconstruction in the American South*, American Ways Series (Chicago: Ivan R. Dee, 2007).

11. Spencer C. Tucker, *Brigadier General John D. Imboden: Confederate Commander in the Shenandoah* (Lexington, KY: University Press of Kentucky, 2003).

12. Sean Patrick Adams, *Old Dominion, Industrial Commonwealth: Coal, Politics, and Economy in Antebellum America* (Baltimore: Johns Hopkins University Press, 2004), 1–3.

13. Ibid, 10–11.

14. Tucker, *Brigadier General John D. Imboden*, 294.

15. Ibid.

16. Ibid.

17. Ibid.

18. Ibid.

19. Ibid.

20. Whose brother surveyed the Pennsylvania Coalfields.

21. Adams, *Old Dominion*, 125.

22. Rogers' survey ignored for its realities in Pennsylvania—according to Anthony Wallace.

23. Adams, *Old Dominion*, 122–51.

24. Ibid, 152–81.

25. Ibid, 189–222.

26. Tucker, *Brigadier General John D. Imboden.*

27. Michael Fitzgerald, *Splendid Failure.* Virginia's Conservative Party was against Reconstruction.

28. Hotchkiss, Jedediah, Volume 1.

29. Did not happen, though the population did grow. Refer to the Talmage Stanley in the Introduction on local county growth in Virginia.

30. Hotchkiss, Jedediah, 22.

31. Ibid, 2.

32. Ibid, 5.

33. Ibid, 21.

34. Ibid, 26.

35. Ibid, 40.

36. Ibid, 32.

37. For example, according to STS scholar Travis Williams, Silicon Valley overly determines as a place of net positive technology and environmental care and

adherence to regulation. However, the actual materials and workers building Silicon Valley-designed products likewise become black-boxed, hidden behind the narrative of progress.

38. Thomas Kuhn. *Structure of Scientific Revolutions* (Chicago: University of Chicago Press, 1962); Robert K. Merton, *The Sociology of Science: Theoretical and Empirical Investigations* (Chicago: University of Chicago Press, 1972).

39. Mary Richie McGuire, "The Chesapeake: Bacon's Promise, Boyle's Project' in 'Translating Natural Knowledge in the Age of Revolution: Tobacco, Science, and the Rights of Man and Nature in the Journals of Benjamin Henry Latrobe, 1795–1820" (PhD Diss., Virginia Tech, 2018); Michel Foucault, *Society, Territory, Population. Lectures at the Collège de France, 1977–78* (London: Picador, 2007); "We make the large knowable by reducing it to charts, maps, and so on. Latour circles back to a world made of paper." Steven Shapin and Simon Schaeffer, *Leviathan and the Air-Pump: Hobbes, Boyle, and the Experimental Life* (Princeton University Press, 1985), 541; Bruno Latour, From Texts to Things: A Showdown: Nature is not directly beneath the scientific article; it is there indirectly at best (laboratory provides the paper with the visual display) (67) in Kaplan, David. "Readings in the Philosophy of Technology." (Lanham, MD: Rowman & Littlefield Publishers, 2009).

40. Foucault, *Society, Territory, Population.*

41. Shapin and Schaeffer. *Leviathan and the Air-Pump*; "National Security Strategy 2010." Administration: Barack Obama. Executive Branch, n.d.

42. Carolyn Merchant. *The Death of Nature: Women Ecology, and the Scientific Revolution* (New York: Harper and Row, 1983), 2.

43. Ibid, 3.

44. Ibid, 37.

45. Ibid, 38.

46. Rebecca Scott, *Removing Mountains:* 2010.

47. Ibid.

48. Deep mining can leave a relatively small surface print. The contamination from deep mining, however, can come with onsite chemicals used to clean or process the coal and with dust during loading and/or water passing through the mine and the site. By contrast, surface mining can involve thousands of acres. A common practice in the United States in surface mining until the 1990s was to remove topsoil and simply push it off the side of the mountain. Thus, once the coal had been scraped, a lifeless, biologically inert "soil" was left for reclamation. The common practice was also to compact the surface, which made the space highly susceptible to fast-moving runoff during rain, allowing for increased flooding as the water rushed, unhindered, toward the places below. Moreover, the entire area of a surface mine was off limits for locals and could cut off access to burial sites, homesteads, hunting grounds, and so on.

49. James C. Scott, *Seeing Like A State: How Certain Schemes to Improve the Human Condition Have Failed* (New Haven, CT: Yale University Press, 1998).

50. Wallace, *Saint Clair.*1987.

51. Sheila Jasanoff, *The Fifth Branch: Science Advisers as Policymakers* (Cambridge, MA: 2009), 65.

52. Ibid, 5–11.

53. Ulrich Beck, *Risk Society: Towards a New Modernity* (London: Sage Publications, 1992); Jerome Ravetz, "The Post-Normal Science of Precaution." *Water Science and Technology* 52, no. 5 (September 2005): 11–17; James C. Scott, *Seeing Like a State*, Ch. 8 and Ch. 9.

54. James C. Scott, *Seeing Like a State*, 309.

55. Travis L. Williams, "Corporate Accountability and Environmental Health Advocacy in Silicon Valley" (Paper presentation, Dimensions of Political Ecology Conference, Lexington, Kentucky, February 2014).

56. Ibid, 2014.

57. Author's field observations, 2015.

58. Travis, "Corporate Accountability."

59. Bruno Latour, *Science in Action: How to Follow Scientists and Engineers through Society* (Cambridge, MA: Harvard University Press, 1987).

60. Joerges, Bernward, and Lars Ingelstam. "Large Technical Systems and the Discourse of Complexity." Complex Technical Systems. Stockholm: Affärs Litteratur, 1996.

61. Ibid; Mike Hulme, *Why We Disagree about Climate Change* (New York: Cambridge University Press, 2009).

62. Much has been made of the need to buy local, be local, develop local as popularized through the works of Michael Shuman's book *The Small Mart Revolution: How Local Businesses Are Beating the Global Competition*. The issue becomes that single sector rural places once overly developed in population and infrastructure and that remain well outside a commutershed cannot compete with small towns and rural spaces within commutersheds. Likewise, many of these rural industrial spaces are left holding the bag on environmental cleanup, pressing medical and other needs, and have neither leadership nor the numbers to compete with urban areas for federal grants or investment. Residents largely must export product if they wish to stay local as there are not enough local buyers to keep the economy afloat much less robust.

63. The philosopher of science Paul Feyerabend downgraded the importance of empirical arguments by suggesting that aesthetic criteria, personal whims and social factors have a far more decisive role in the history of science than rationalist or empiricist historiography would indicate. Paul Feyerabend, *Against Method: Outline of an Anarchistic Theory of Knowledge*. 4th ed. (New York: Verso, 2010).

To this, I assert Michel Foucault: if the economic take-off of the West began with the techniques that made possible the accumulation of capital, it might perhaps be said that the methods for administering the accumulation of men made possible a political take-off in relation to the traditional, ritual, costly, violent forms of power, which soon fell into disuse and were superseded by a subtle, calculated technology of subjection. It would not have been possible without the growth of an apparatus of production capable of both sustaining them and using them . . . conversely, the techniques that made the cumulative multiplicity of men useful accelerated the accumulation of capital. Michel Foucault, *Discipline and Punish: The Birth of the Prison* (New York: Vintage Books, 1977).

64. Michael J. Mulkay, "Norms and Ideology in Science." *Social Science Information* 15, no. 4–5 (1976).

65. Author's field observations, 2015–2016.

66. Mulkay, "Norms."

67. The World Bank, 2010.

68. Information about Massey Energy is presented on the Center for Media and Democracy's SourceWatch.org entry at https://www.sourcewatch.org/index.php/Massey_Energy:

> Massey Energy describes itself as "the largest producer of Central Appalachian coal and America's 4th largest producer of coal by revenues." In 2009, the company "sold 36.7 million tons of produced coal generating produced coal revenues of $2,318.5 million. [1] In 2008, the company had 6,743 employees. [2] Of these, only 1.3 percent—or approximately 87—were represented by the United Mine Workers of America. Massey states that union members were "spread out amongst five of our coal preparation plants" which handled "approximately 15.8 percent of our coal production." However, the company states that the "collective bargaining agreements with the UMWA have expired" and that "there are no ongoing negotiations" at present.[1]

In 2008, Massey estimated that it controls 2.2 billion tons of coal reserves in southern West Virginia, eastern Kentucky, southwest Virginia and Tennessee, or almost one-third of the total coal reserves in Central Appalachia.[3]

For details on the April 2010 Upper Big Branch Mine Disaster see the Upper Big Branch Mine Disaster article also on *SourceWatch.*

In 2011 Massey Energy was acquired by Alpha Natural Resources which is a major donor to Virginia Tech. See Appendix A.

For a true sojourn into the absurd politics of this region, the federally indicted former head of Massey Energy, Don Blankenship, made an unsuccessful bid for U.S. Senate. See: Janet Hook and Kris Maher, "Don Blankenship, Former Coal CEO, to Run for West Virginia Senate," Wall Street Journal, last updated November 29, 2017, https://www.wsj.com/articles/former-coal-executive-convicted-on-charges-tied-to-mine-explosion-to-run-for-senate-1511987189

69. Dan Barry, "As the Mountaintops Falls, Coal Town Vanishes," *New York Times*, April 12, 2011. https://www.nytimes.com/2011/04/13/us/13lindytown.html

70. Gabrielle Hecht, "Rupture Talk in the Nuclear Age: Conjugating Colonial Power in Africa." *Social Studies of Science* 32, no. 5/6 (2002): 691–727.

71. Achille Mbembe, "Necropolitics." *Public Culture* 15, no. 1 (Winter 2003): 11–40. The source for the contemporary conversation on necropower. Here I examine Mbembe's necropower as it applies to the dead physical human body for its conceptual application to the dead or murdered town. In essence, I have started with what may be a set of extremes: a monotown at its most extreme as analyzed by a critical theory examining social relations at their most extreme. This move affords me wide critical room to later move toward more moderate monotown examples and critical approaches.

72. Ibid.

73. Ibid, 23.

74. Ibid, 16.

75. Draglines are large excavating machines used in surface mining.

76. Flyrock is rock expelled from rock blasting.

77. Letson, Al. "Appalachia Rising." *States of the Re:Union*, 2011. (Site discontinued. Episode not present on residual YouTube channel.)

78. Mbembe, "Necropolitics," 11.

79. Ibid, 12–23.

80. Emily Satterwhite, *Dear Appalachia: Readers, Identity and Popular Fiction since 1878* (Lexington, KY: University Press of Kentucky, 2011).

81. Mbembe, "Necropolitics," 36–37.

82. Ibid, 39.

83. Ibid, 38.

84. Maybe this is an odd place for this aside , but popular culture even has its homages to this loss of place. While writing this book I often thought of the lyrics to The Pretenders' "My City Was Gone" (1982). In this song, big box consumer culture has replaced a different conception or hometown:

> I went back to Ohio
> But my city was gone
> There was no train station
> There was no downtown

This process comes less as news post-1980s. Billy Joel's lyrics speak more to this failure of the American "way" as THE American Way. Released the same year as the song above (1982), some lyrics from "Allentown":

> Well we're living here in Allentown
> And they're closing all the factories down
> Out in Bethlehem they're killing time
> Filling out forms
> Standing in line
> And it's getting very hard to stay

85. Mbembe, "Necropolitics," 14.

86. Karl R. Popper, *The Open Society and Its Enemies, Vol. 1, The Spell of Plato* (Princeton, NJ: Princeton University Press, 1966); Popper, *The Open Society and Its Enemies, Vol. 2, The High Tide of Prophecy: Hegel, Marx, and the Aftermath* (Princeton, NJ: Princeton University Press, 1966); Yaron Ezrahi, *The Descent of Icarus: Science and the Transformation of Contemporary Democracy* (Cambridge, MA: Harvard University Press, 1990).

87. Ezrahi, Yaron. *The Descent of Icarus: Science and the Transformation of Contemporary Democracy* (Cambridge, MA: Harvard University Press, 1990).

88. Such as Environmental Science or Reclamation Science.

89. Ezrahi, *Descent of Icarus*, 51.

90. Beck, *Science, the Endless Frontier.*

91. Ezrahi, *Descent of Icarus*, 61.

92. Popper, *The Open Society and Its Enemies*, Vol. 2, 158, 162.

Interlude 2

Atlas Charles

To arrive at conceptualizing these interludes as both necessary and scholarly, I am forever indebted to Myles Horton's and Paulo Freire's *We Make the Road by Walking*: *Conversations on Education and Social Change*.[1]

The summary of this immensely accessible volume runs like this:

American Civil Rights activist and Highlander Center cofounder Myles Horton argues the "system" is rotten, and you must make changes outside it, and replace the system.

The Brazilian advocate of creating a critical consciousness so that people can change and challenge the circumstances of their lives, Paolo Freire advocates for making change within the system, because, as I thought my way to later, there is only the system. There is no truly outside society, and in colloquial American, you need to dance with the one who brung you.

In 2016 in response to input from respondent interviews with grasstops, I piloted three working groups for intra-regional work based in this subregion of Appalachia not limited to but also encompassing the Pocahontas Coalfield. In the end only one of these working groups took off—one focused on agriculture. This was not entirely surprising as this was the region's main sector prior to energy sector development, and, it remained a continuous though uneven sector even during the heyday of coal employment.

In contrast to these working groups, writing an academic book stands as one of the last "solitary" acts of scholarly creation. Many fields now demand collaborative research and scholarship. I could have stopped this volume at the account of how the Pocahontas Coalfield happened, and, its description as a type in Big Rural.

Yet, I could not rest in peace with the metaphorical flattening of the rural space as described by James C. Scott. These interludes pull that thread of history and reflection on through to the now: I cannot rest in peace with the flattening of democracy and the physicality of lives in the Pocahontas Coalfield.

Oral history as a means of historical and literary scholarship has a strong history in Rural, Appalachian, and Regional Studies. Approaching the recent past, oral history affords the people involved in making that past with the possibility of recounting their own history in their own words. As a practitioner in the region both as a scholar and a person involved in making sector change, I have a responsibility to make space for voices, perspectives, stories, and theories of people in the region in their own words.[2]

In the spring of 2017, I became acquainted with Atlas Charles at the Appalachian Studies Conference. Atlas offered that they were from Buchanan County, VA (also a coalfield county) and had lived for a while in McDowell County, WV. Atlas has since skyrocketed to be a leader in social change and regenerative economic development in the region, mixing a blend of radical pragmatism regarding the nuts and bolts of municipal, nonprofit, and business stewardship, and, regional, municipal, and nonprofit envisioning.

Atlas Charles (*they*, *them*, *their*).

Atlas was speaking to me from their home in Bristol, TN.

Crystal: I came originally out of fiction, and I know that you are also a storyteller. You and I share a belief that narrative is important. How did you get here?

Atlas: I am going to back up some conversations we were having before this about disinvestment by the State and imaging the future of places. We can't get to a place of regenerative potential, where we are nested in the world from the earth down to the environment and the economy, because the way we think now is that the economy consumes everything.

Part of the reasons the state has abandoned places like McDowell County, or abandoned the common folk, is because the paradigm is flipped. Not only are we dreaming of new futures, but we are building new paradigms for how those systems work. The system we are working under now is fragmented, if you make an intervention over here or over here, and you're still going to get some good impacts. It's all based on the disciplinary higher-education model of how knowledge is thought to be passed on.

In my coaching work with organizations, the narrative is often where I start. I want people to tell me their story, even if people don't conceptualize their life as a story. The reason I think it is important that you are bringing the narrative and personal story into this work is that I am a community wealth advocate because of my personal experiences. The truth is I am good at my job not so much because of my professional experience, but because of my homegrown systems thinking. This came from how I grew up, where I grew up, and figuring out how I was located in those systems. Narrative is the bedrock of that.

Crystal: So, who told the stories in your family and how did you come to see yourself in the story?

Atlas: I think my grandma, and, my granny. They weren't story spinners, but they were community weavers. In that community, there were stories that emerged. People would come in and out of my granny's house to eat because they knew they could get a meal there. We lived in a holler and somebody would be walking by, and she'd be trying to feed them.

In a lot of ways, my granny held our family together. She was this holder of connection and connecting people. My family is pretty strongly matriarchal, although my dad never would have admitted that. My grandma is the one my family goes to for wisdom, stories, and care. On my dad's side, she has been the anchor since my granny died.

On my mother's side, I don't remember my grandma much before she got rheumatoid arthritis and other illnesses. She developed that pretty quickly after we lost our house in a flood in 2001. Growing up in a holler with a family that weaved community, although the relationship with nature had already largely been broken, I was taught in some ways to be in community, to be with nature. That is where all the work was started because that is where I was taught to care for those outside of me.

My granny knew that she could feed people and she'd have what she needed the next day. There was a complication though with my own weaving into that community as a kid. The most obvious one was growing up as an autistic kid in that community. Autistic kids always feel out of place and like they don't belong. I have this very clear memory of being a kid and being in this crowd of family, and I felt like a fish out of water.

Thirty years ago there wasn't this diagnosis, so I was just this weird kid who rolled instead of crawled and was a little bit too smart. My family just sort of accepted that. On top of that, I knew when I was four that I was queer. I oppressed that until I was sixteen or so. I also had the out of placeness of being queer in a place where you know that is not going to be accepted or nurtured. There weren't many models. My home was riddled with domestic violence. My earliest memories are of fights. We lived with my dad till we were seven, and then my mom moved. This was also a really hard decision for her, because her family had its faults, too. She has talked to me about how difficult that was to do because going back to her family, she also knew the problems there intimately.

The way I talk about all this now is that I was deeply connected to the place and the people there and the ways of life, in Buchanan County (Virginia). I have connected also to the animals around me. I would go outside and make things out of berries and twigs, climb chicken cages, and want to help my dad take care of the animals and the garden. I wanted to go play in the woods. The holler that I grew up in and the land that we lived on has been my family's since the 1760s or so. We predate the mines. My German ancestors that settled there farmed, and they were a rather wealthy family in farming for years. Then that wealth just plummeted during the coal years.

I was surrounded by my granny right down the hill. I lived with my mom and dad. My papaw lived with my granny. My grandma, for part of my childhood, lived just up the road. My uncle lived there. I knew the kids that lived across from them. We had some more distant relatives on up the holler. There were quite a few people I grew up around and felt safe around.

I would say that my parents did not share the bulk of the burden of imprinting on me because they were working a lot. My granny and I were really close. Another one of the reasons I feel out of place around people now is that I grew up with an eighty-some-year-old granny as the main imprint. My dad's family had a history of Appalachian granny witches. Granny witches were the ones who did the medicinal healing from natural things or took care of people in death or were midwives. Often just general caretakers of the community in different ways. They drew on the traditions of the community. I mention that because I struggle now with how devoutly Christian my dad's family has become. They used to have more connection to nature than the sterile Christianity promoted now.

Crystal: On this topic, I was thinking the other day that every holler had its church, its people, its community, its relationship to nature, and, to your point, that just as television and the larger connected society now have pushed a more standardized way of speaking, it would be surprising if churches and what is expected at them and in them had not also standardized. Just as what evangelists on TV push as Christians, TV and other media push what is supposedly normal. My family was Presbyterian, and they certainly considered themselves believers, but often they were too busy to go to church. And, there would have been room for your granny to heal and it wouldn't have seemed un-Christian.

Atlas: My granny belonged to the Old Regular Baptist Church, not a large denomination. In this church people let themselves show emotions in church, like getting happy, but not charismatic. You did not see a lot of emotion from these same people otherwise. There are also a lot of lay ministers.

I learned pretty quickly as a kid not to want to go back to that church as there were certain things I could not reconcile with their beliefs. There has been an increase in polarization in the last twenty years along religious and political lines, and this has even affected the relationship with nature, this broken relationship with nature. In any case, I find nature now a large part of my identity, to have this family that has had a strong tradition of Appalachian granny witchery; it is not paganism or witchcraft. There is a reliance on nature. I feel very connected to the land.

I feel very disconnected right now, in a city, with a small backyard. I grieve a lot for space, for quiet, for mountains I can live with and alongside.

I have been struggling with being in a place where other facets of my identity, being queer, autistic, and in the trans community, and I can live as me, but how in the hell do I get connected to the place where I lived? I can never go back to that holler and have these things, but I also can't have what I deeply yearn for, a connection to nature and community in those same ways in the place where I live now.

Crystal: Yeah, I don't want to be the person that feeds you drivel. You know, the people who say, well, you can just have it all. Just go have it all! You know, like folks did in the 1970s where they went and built their fortress communities in the woods. Other folks don't have to think about or do that. They can just go about their lives, and if they have the resources, have their second homes or cabins or whatever in the woods, and live their lives otherwise in the city and going back and forth.

Atlas: Yeah, I use all my mental energy to live my life where I am right now much less trying to live two separate lives in two places. You can't have it all. That is an important reality for queer Appalachians that we face that people need to understand. We are forced to choose between our family, our place, our environment—things we deeply care about, but we have to leave in order to take care of our mental health and to take care of ourselves.

You know, in small towns, you can't exactly piss off your neighbor every day because you are going to see them at the grocery store. People don't tell me that I am evil and going to burn in hell for being queer, but they think it. In cities, people are more direct, but they care less about that kind of personal level stuff.

In the small town where I grew up, I am "their" queer. As a queer human, I was rather well-accepted on some levels as I owned a business back home. I have been involved on and off in the community since I left twelve years ago. I am the "good" queer and I have intentionally played that role—the good citizen. It has kept me safer in some ways but also kept my mom safe. She would have had also to deal with this fallout.

Part of my story, and why I do the work that I do, is because I have had to choose between this place and these people that I truly and deeply and intimately love and value, and being in a place where I could be seen and safer than what I would have back home. Even though people back home don't take out the pitchforks or egg my house, it doesn't mean that they see me and don't deadname me. This is a large part of my story because if I could have both of those things, I would be back home. This is part of the reason I do the work that I do—

Crystal: So, let's turn now to the work that you do.

Atlas: The work that I do in community wealth building, the reason I do that work is so that kids like me can thrive. I want a Hurley, Virginia, and a holler, where a queer kid is not going to have to make a decision between a place and the people they love and deeply care about and leaving to get a job, or to be in a place where they feel safe and seen in, that they feel like they can live authentically in. So, that's the way of the work.

I have had to choose, and I don't want that for people. When you grabbed me at that conference, we talked and connected, and part of that was just being where we were from. I showed up in Bluefield, WV when we were originally going to do a project there, and I remember leaving this meeting we had there, and I had written a business plan for a burger and chicken joint using local food as part of a planned complex. I had a pizzeria at that time in Hurley, VA so it made sense for me to write that, but I remember enjoying myself in a way I had never enjoyed myself with work before.

Writing that business plan, I knew somewhere deep inside me that this is what I want to do for the rest of my life: plan businesses and help create them in small towns. I was like, how do I get to do that? I had some inkling of this in undergrad when I had written a paper critiquing the ethics of economic development in Appalachia. The critique started with first recognizing the harm that had been done: logging, mining, and other extractive industries, not only to the land but by erasing knowledge, heritage skills, and what these had left in their wake.

In that paper, I looked at creativity and dreamed of a post-coal future. This was 2010 before this became a more mainstream conversation.

Crystal: That was about the same time I started my Ph.D. I had a similar question. How do we do this?

Atlas: Yeah, we were not going to have an Appalachian Moses pop up and lead us out.

Crystal: At that time, if I went to conferences, the discussion was about some magical industry or factory or company that Appalachia was going to attract in. Some magical outsiders were going to move in. I scared some people to death in about 2011 at a conference on women and leadership by asking—what if this is it? These are the people we got. These are the resources we have. What if the checks, federal, personal, stop coming, then what? It's not improbable, and it's already happened. There is never going to be the devotion of resources to "fix" all that has been wrought, yet, we are not resourceless ourselves. Not just coal, or minerals. We are sitting on resources that people can use from everything for a livelihood to an economic sector.

Atlas: I would call those resources "wealths." At the time of my undergrad, that is when I also started looking at food, as small-scale farming is what we did pre-coal, and not that we go back exactly to that, but the other industries that could be attracted in seemed to offer the same result that coal has.

On my way to this, I worked in Arkansas and did community work, and saw firsthand the school to prison pipeline of the Black community there. I did a lot of work there on my own paradigms on diversity, equity, inclusion, and by the time I came back to grad school in Appalachia, I was no longer believing that my poor White ass was the worst of the worst. I knew that my Whiteness had a lot of privilege. I had never been taught to recognize it, and it was hard to recognize.

Crystal: I often think of this as land or object wealth. Even after the Civil War, even landless White people were able to take with them some fruits of their labor which still translates into some reason that I sit here today. That was the land my grandfather was able to buy mid-twentieth century and my dad could get loans to turn that into other wealth.

Atlas: The reason my family had not left Buchanan County, VA, at least as late as the 1970s, was they still farmed along with working in the mines. However, there has been a terrible introduction of substance abuse in my family that parallels the introduction of the mines. We think of the disappearance of the mines as creating the addiction issues we see now. I have intimate knowledge of the complexities, where the very mine my dad worked in on the mountain where we lived also polluted our drinking water.

People forget that mining is hard on your body, and that could create substance abuse disorders. But what we have now, we have heroin, fentanyl, and meth, those are just harder drugs. Before that was alcoholism.

Crystal: Yeah, my great-grandfather, who did railroad work and worked in the mines, when my great-grandmother was asked what he died of, she said, "the bottle." My grandfather died of complications of black lung and emphysema right about the time that a benefit was introduced. He left the land he had worked for and bought, so my grandmother was land rich and cash poor, so my dad went to work along with going to high school.

Atlas: Arkansas was a deep lesson, and I came back to Appalachia with a different perspective.

Crystal: Yeah, it's not like we are getting special ill treatment here in Appalachia. What happens here happens all across the country, all across the world, and this sense that Appalachia is somehow singled out for it keeps us from having solidarity and working with other similarly affected regions.

Atlas: I saw more of "Appalachia" there in the Delta than anywhere else I had been. I went to the Southwest and saw it there, where there was disinvestment and disenfranchisement. There is a lot of power in figuring out how Appalachia is similar to other regions rather than how it is different.

But back to graduate school in Appalachia, out of that, I got chosen to take a class and work with the Appalachian Regional Commission to engage in economic development oral history in forty communities. What I took away from this was that people write a lot of bunk in their grant reports, people do a lot of nonsense with grant money, which includes getting grant money to assess a problem that you already know is a problem, and then that problem never gets addressed with your grant money.

The third thing I walked away with was that economic development in the United States uses a lot of disconnected metrics to inform whether the economic development is going well or not. So, something like jobs created or jobs retained. Those are the two that Congress cares about. The truth is that these two metrics tell you nothing about people's well-being, the availability of public transport, the quality of education, the things that would actually give people reasons to live in a place.

My biggest critique is that these jobs created and jobs retained are abstract, and not tied to reality or community. The bottom line is what ought to happen is making people's lives better. I know what it is to want to die. I spent over twenty-eight years of my life wanting to die with suicidal ideation, and I did not want to exist. For two years now, I have had only two thoughts of suicidality. By my mid-twenties, I had to get very clear about what it was that I woke up for every day.

At the very core of that is I have to have joy. I have to have community. I have to have people that I love. I have to have work that I love, that I feel is making a difference. Our world is burning. People are dying. And if I can't find the courage to do work and live and love in ways that get us to a better place, even if just one step closer before I die, then I didn't really find any meaning in living. Building communities where kids like me can thrive means building communities where everybody can thrive.

Crystal: I cover this elsewhere in the book, that when I went to conferences on economic development in the region to observe as a scholar, I saw the same types of people. It seems this has shifted, at least in lip service, but if the checks don't get written, then I don't believe the shift is real.1

Atlas: It's performance, and it's even more harmful than having no policy. It's basically gaslighting people. We have a policy that protects you or funds you but we are not actually going to protect or fund you. In the last two years, particularly since the murder of George Floyd, and also the murder of Breonna Taylor, which is closer to home here, there has been an acknowledgment of how resources need to get to the ground, but overall, I have not seen a huge movement toward pushing out resources to Othered people.

It's still like, it's on your own bootstraps. You just didn't pull them tight enough. But my day job is helping organizations with strategic planning around community wealth planning and regenerative economics. My message is really, at the end of the day, to foundations or wherever—let's take McDowell County for example. The disenfranchisement, the barriers that people face, the overall sense of hopelessness, folks there have been beat down, beat down, beat down.

There is not intentional support to people in those kinds of communities, to write whatever grant to jump through whatever hoops the federal agencies put up in their request for proposals (RFPs). These federal agencies need programs to help places like McDowell and the already strained grasstops in them to jump through those hoops. Or just take the hoops out. That would be the better way.

Crystal: One of the pieces missing in all this and you have brought up these federal metrics—the issue is that the metrics are all external. Communities themselves ought to come up with the metrics. If I am in a community and making change, it's not that we meet your metrics, it's that we come up with what a sustainable community means to us.

Imagine an economic proposal to a federal agency, but also sent to similar people in places like McDowell County, WV. On this project, what would succeeding be for you? What if the community designed the metrics? What may be mundane in one community could be revolutionary in another. Communities in the same straits may recognize that in a way that federal agencies expect homogenization, and run their metrics based on a flattening of place. Community-based metrics must be a shift that happens. In the meantime, at least we can write our own stories and what ought to happen.

Atlas: Let's take the bioregion that Appalachia sits in—or maybe between the Shenandoah and the Ohio River down to the plateaus of Tennessee across to I-77 in North Carolina. Imagine if corporations even in that region were taxed at a rate that they should have been and that money was reinvested in that region.

One addition to this triple bottom line that people talk about—people, planet, profit—I add place. If you think about where I am from, Buchanan County, Virginia, if the corporations that made money there had been taxed accordingly and that money put back into that community, Buchanan County would have ended up one hell of a great place to live.

Crystal: One part of the project of this book is that we invent a place. The Pocahontas Coalfield was invented. These inventions do not last necessarily.

Atlas: Not only did people invent the coalfields, but we are actively inventing places right now. Place is many different things: the community, the environment, the future of the place. Well-being is at the center of that.

For me, neither capitalism nor communism will lead us out. We need a new regenerative economy where not everything is a transaction. Socialism is an economy in service to the State, not in service to the community. A market-driven economy works but not in the capitalist system we have set up. Local economies need more autonomy, but I am not so naïve to think globalism is also not still in the picture here.

Crystal: By contrast, our autonomy has been outsourced, our democratic agency is outsourced, if we do not have the control over the measures, the wealths, to enact or live democratically where more people are freer, there is more equity, there is more accountability. Federal agencies, foundations, nonprofits, corporations, universities decide what is important for us rather than us.

Atlas: Yeah, these groups are not our Moses either. . . . Coming full circle, this analysis, this work on community wealth, all of that is fed by growing up in Buchanan County, VA, and McDowell County, WV. And at the end of the day, I believe it is those personal experiences, as an Appalachian, as trans, as queer, and as autistic that weave into those professional responsibilities to make what I do effective as a strategist and facilitator. That is what I invite people to do a lot. Bring your personal experience in. You are the one who has the passion. You are the one that has the creativity, the innovation, the ideas. It is not your professional persona that keeps those things. It is the heart. You have to work through that messiness with yourself before you can show up with that for other people.

Crystal: And some of that mess is the situation—the current drugs, the current addictions. Those are systems and you can't work through those on your own. And more people don't escape that than do.

Atlas: Yeah, I am thirty years old, and I have lost thirty-six people to addiction, or suicide, which is mostly in the queer community, or some other poverty-related death. At the end of the day, how can I stand by and not do the work that I am doing? People are dying. People that I love. My dad died two years ago from an overdose of fentanyl. My adopted grandpa overdosed. Numerous friends—I wake up and they're gone. There is no way that I can't do the work. Imagine by the time I am sixty I will have lost over 100 people in my life. This has all doubled in the last five years.

Part of that is I have got to do the work of dealing with that trauma. I have to show up in my professional work dealing with the trauma of losing that many people or dealing with the trauma of my childhood. To show up as my full self, I have had to make sense of my own story. The work I do now is like you have to be able to hold tension, like when my dad was a kingpin [drug dealer]. Or when my dad was a coal miner. That the money feeding me might not have been the "best" money. It was money that was harming people, ultimately. But it was money that fed the family. The complexities that people live in Appalachia are what I try to go to work with.

You have to balance economic growth with well-being. That well-being takes precedence in that. Coal mines mean black lung and injured workers, but no mines mean people lose their jobs and my sister and her family might not have enough money to eat.

Crystal: I mean, the mines put me here today. My grandfather worked in them; that money leveraged into a different reality for my dad, and an even more different reality for me. My grandpappy could only go to the sixth grade as that was all there was and then he went to work as a water boy for the Italian stonemasons working in the coalfields and worked around the mines after that. My dad had a support business for mining for heavy equipment. I am standing on their shoulders. We also have to be real that the mines had also made people middle class, and then it also did a whole bunch of stuff that was not that.

Atlas: I am the first person in my immediate family to go to college. My mom had the foresight on that. I know this is not always the case. It democratized my life in that way. My work is walking hand in hand with local people in local places to create anti-oppressive economies, and economies that regenerate rather than extract. Without that complex understanding is when solutions become problems.

Crystal: And if we don't do this, then who? If not now, then when?

Atlas: If not now, then how many more people die?

Crystal: There are people that walk away and choose not to think about it. By contrast, empathy is an action.

Atlas: And that I am going to sit beside you in the pain. I know we are not the only ones talking about this. The thing is that I am coming to with place, the throughline that I am coming to with this work is that it's about the place, it's about their holler, their immediate place that taught them to love and to be in community. I am finding more and more that it's about the holler; that it's not about Appalachia overall. The blanket of Appalachia covers us all to outsiders, but when I talk about home to people in Appalachia, they want to know where specifically. That's maybe my new term: it's about the holler.

Crystal: And the holler could be your Brooklyn neighborhood. You are from around the way. You are Jenny from the block. What are we really actually wired for? Maybe we are wired for the "hood", we are wired for the holler.

Atlas: That is where I am coming to. That is the yearning that I was talking about. I yearn for the 100 people of that holler, and the land, and the animals, and the nature that were in that holler.

Crystal: We are working on creating our holler intentionally.

Atlas: To create the holler where other people have a chance in , and since I have the ability to do this work, I have the responsibility to do it. Eventually, I want to be able to live in two places.

Crystal: You can't force community—

Atlas: But you can intentionally create and cultivate and hold that space for the community to develop. Forcing the community is antidemocratic. But like coming together as a community and in service to the common good. But not forcing folks together like communism or apart, like capitalism.

NOTES

1. Myles Horton, Paulo Freire, Brenda Bell, John Gaventa, and John M. Peters, *We Make the Road by Walking*: *Conversations on Education and Social Change* (Temple University Press, 1990).
2. Lynn Abrams, *Oral History Theory, Second Edition* (Routledge, 2016).

Chapter 3

Democratic Possibilities and Policies in Big Rural

OVERVIEW

Liberty and equity ought to be cornerstone ideals of both a functioning liberal democracy and of a functioning policy in a liberal democracy. I demonstrate this argument by proposing an approach, a strategy, to inform science and technology practice and research in and on the rural industrial. I also agree with political scientist Deborah Stone that liberty and equity have material and real applications in the polity; they can be enacted and practiced. Likewise, people also create the economy and the market—none of these constructs (humanly created concepts) function through immutable laws or absolutes; they all must be defined, redefined, and exercised to attain material substance.[1] All are works in progress, like my conceptualization of Big Rural, and their characters and abundances are determined by real human actors in society.

The basis of policy in a democratic state ought to be to increase liberty and equity for individuals and communities—as they have the least organized protections in the United States in comparison with a range of corporate entities. In the application of policy, I concur with Deborah Stone, that in the American context, corporations have very few restrictions on the consequences of the actions of their agents. She explains that "public policy must address conflicts between the liberties and interests of individuals and those of corporate actors" such as "churches, trade unions, sports franchises, professional associations, business corporations," and so forth. Because these entities can exhibit such control over individuals, their power ought to be especially limited.

In the United States, we falsely assert that these actors and private corporate actors are "weak" and that the individual is strong, when by contrast, these kinds of groups assert tremendous power over workers, the fate of communities, and over society.[2] Rather than a conflict between equity and liberty, or liberty as construed as the absence of harm, another way to consider liberty is "whenever a person's control over . . . [their] life is increased." Stone prescribes power (such as voting and meaningful social decision making), wealth (material resources), and knowledge (capacity to envision "solutions" to "problems") as necessary to increasing liberty and as measurable. Liberty is available by amount, by degree, and can be maximized. How much liberty one has relates to how able and amenable a society remains to addressing the attributes of liberty issues directly.[3] I add here my reluctance to speak generally of society; by contrast, I have named specific people and classes/categories of people who must be amenable or be forced, through direct exercises of democracy such as bottom line-affecting strikes, to be amenable.

With respect to equity, Stone highlights that every issue of policy accords an issue of dispensation, whether material or of making one's luck or having one's luck dictated such as compulsory military service, access to education, or holding an elected office. Rather than go through her brilliant analysis of the complexity of defining equity (worth reviewing if not familiar), more important are the questions she assigns to the policy analyst. I paraphrase Stone, indicated by parentheses:

(Stone) First, who are the recipients and what are the many ways of defining them?

I have spent considerable time in *Big Rural* defining the macro and micro of the group at issue, the people of the Pocahontas Coalfield currently aka micro and people in a rural industrial single-sector region, and in Big Rural more generally aka macro.

(Stone) Second, what is being distributed and what are the many ways of defining it?

My examination, and answer to this question: the practice and benefits of equity and liberty available in the Pocahontas Coalfield and in the context of a rural industrial space currently in the United States, and, in other similar spaces elsewhere.

(Stone) And third, what are the social processes by which distribution is determined?

I have reviewed some of the social processes at stake in the Pocahontas Coalfield and in similar single-sector rural industrial spaces:

- Work and town conflating into the totalitarian state. The Pocahontas Coalfield predicts what happens in a single-sector rural industrial space.
- Science and research serving corporate liberty rather than individual or community liberty. Corporate science ought to be differentiated from science for the sake of producing knowledge.

- Corporations automating away workers–citizens or making a space uninhabitable for citizens.
- Corruption by state stakeholders on behalf of themselves and corporations.

POLICIES

In response to the fulfillment of actionable liberty in the Pocahontas Coalfield, I have conceived of policies incorporating informed possibilities for potentially increasing liberty and democratic practice specifically there and in similar spaces where technology and corporations have especially negated metis, environment, and liberty.

During interviews for *Big Rural,* respondents surprised me by listing lack of vision and lack of leadership as the top issues in their regions (Pocahontas Coalfield and coalfield adjacent).

I first discuss a range of approaches and possibilities for reenvisioning the Pocahontas Coalfield and similar spaces. Later in *Big Rural*, I outline actionable rural policy and infer implications for the Pocahontas Coalfield and its case.

DECENTRALIZE LTS AND SUPPORT METIS-FRIENDLY INSTITUTIONS

If we consider the Pocahontas Coalfield as part of the large technical system (LTS) that constitutes energy, then decentralizing this kind of LTS could potentially make energy production both more democratic and less condensed in its associated financial and environmental risks, both for the end-consumer and for people in the current LTS rural industrial space.[4] Even with U.S. production drops in coal-fired power plant-destined coal, coal remains valuable in its use in industry and as an export. Moreover, significant federal R&D spending has remained budgeted for clean coal technology.[5] A long functioning history of more democratic institutions and more metis-amenable institutional structures such as workers' cooperatives and producers' cooperatives allows us to envision how energy production and resource extraction might be reorganized to afford more democracy at the resource site or origin. If coal and energy company administrative structures could be restructured toward more democratic institutions such as cooperatives, worker-owned or citizen-owned LTS, along with the diffusion of technological interventions that place power generation with the end-consumer or even the end-corporate consumer, then it could follow that all of these spaces would increase in liberty; each of these spaces could then absorb more equitably the personal, environmental, economic, and social risks associated with energy production.[6,7]

As I have narrated abundantly, in most respects many of these spaces have already hit bottom as communities (as defined by the World Bank) and environmentally (as I describe in my theorization of Lindytown). To characterize to locals such moves toward democratic structures as "high risk" reeks of a neoconservative assertion that deregulating "invisible hand" market impulses become characterized as high risk. How much more risk ought a place like Lindytown absorb? How much more risk oought Big Rural spaces absorb?

SUPPORT SCIENCE AND ENGINEERING THAT TRULY ADD MORE KNOWLEDGE RATHER THAN FLATTEN METIS: PROMOTE SCIENCE AND TECHNOLOGY THAT SUPPORTS AND STEWARDS THE RURAL

Rather than the flattening of knowledge by Taylorism, and the false assumptions of redistribution of innovation of Schumpeter, *public*-funded science and science and technological education ought to serve first civic purposes such as increasing liberty and buoying democratic practices: in sum, transparency of government, corporate cost-sharing of rehabilitation of land, efficacy of voting (no voter fraud), leadership and civic education, corporate responsibility, and so on. To this list add science that increases worker rights, supports environmental stewardship, and works with citizens to research needed science for economic and environmental sustainability.

Appalachian scholar Talmage Stanley analyzes David Harvey on the flattening of this rural space into a place controllable and controlled by rational material processes:

> Here, Harvey means the abstract and theoretical ways in which we conceptualize and put into language and therefore manipulate the material processes of places: physics, mathematical measurements, social sciences, map-making, geographical concepts, and economic structures. These systems and representations rationalize and legitimate a range of forces and pressures, often condoning systems of injustice and alienation. Whether it be on the level of "common sense" and generally available knowledge, or more abstracted and difficult academic jargon, representations of space are the means of perceiving how society is.[8]

The public universities that currently send mining engineers and geologists to support extractive scientific cum economic practices in the Pocahontas Coalfield—West Virginia University, Virginia Tech, the University of Kentucky, and Penn State, and so on—ought to devote their science and technology education programs to science and technology that increases liberty and democracy as practice for the people in the region where their institutions are located through increased citizen participation in research agendas and from reaching out to the community for research needs. The Department of Mining Engineering or School of Engineering or Science ought to support science and technological expertise to increase democratic practice and more measurable liberty for people in that region, reduce corporate and governmental corruption, and support more democratic leadership in this rural space: those ought to be the stalwarts of energy policy in a democratic society.

These same universities need to incorporate local citizens in citizen science for more democratic land stewardship decisions going forward with the region's assets, and, train citizens in the evaluation of proposed resource utilization.[9] Each science and/or technology department would have to work with local citizens to understand the science at work in the region and the technology proposed before any department could jump on any pro or con bandwagons. This approach could increase knowledge and possibly open dozens of avenues for scientific and technical engagement by scientists and citizens alike.

Moreover, citizens ought to lobby for needed democratic science and technology in their region. Taxpayer funded R&D would need to reflect democratic imperatives. "What of academic freedom?" critics may cry. With the kind of corporate roster currently in play and corporate partnerships for state-funded universities and, with the roster of corporate lobbyists attending to governments, academic freedom would likely increase rather than decrease if citizens became engaged and understood the potential impact of their taxpayer dollars on improving or at least attempting new technical, policy, and scientific approaches to their current problems. Academic integrity would remain integral in "idealized" science and science for knowledge's sake might even return in importance in the United States as it would be seen, given open-source applications now, as a public good.[10]

BUILD ON REGIONAL GROUPS AND THEIR EMERGING COORDINATION

Despite their myriad of economic, health, and environmental problems, neither the Pocahontas Coalfield nor Lindytown finds itself caught in a snow globe, forever swishing in acid mine drainage, decorated in loose rock, peppered with fly ash. Within the area, regionally focused groups such as the Central Appalachian Network, the Appalachian Funders Network, and the Appalachian Regional Commission (ARC) have supported exploration and discussion of other economic possibilities for central and southern Appalachia beyond coal mining. However, directly engaging science and technology's R&D or education agendas has not generally been part of their approaches to change.[11] In particular, the ARC's relationship to the current model for the regional energy sector becomes complicated due to political and other interests involved in their federal partnership. Though the Central Appalachian Network and the Appalachian Funders Network have promoted the concept of "Appalachian Transition," a new discourse in the region, these signify but do not explicitly mention coal. Likewise, the strongest emerging sector, natural gas, largely remains outside discussion when not part of anti-pipeline efforts. Though a powerful industry on the ground, natural gas has not yet attained broader cultural affiliation for this region. Of equal importance, but overshadowed by the environmental and other issues of active coal country, ought to be the issues accompanying any of the region's rural single-sector industries: natural gas in Pennsylvania, Ohio, West Virginia, and Virginia,[12] timber throughout the region, etc. I often think of the *Charleston Gazette*'s section that ran for many years, *Coal Tattoo*, which might have best served the region by also focusing on energy companies beyond coal.[13]

Despite the massive extraction over centuries—timber, minerals, and the ruination of growing capacity (arable land), soil, and water—the region remains rich in a vast array of natural, human, and geographic resources. Various groups in the region seek to work from these remaining assets. Though less connected to these greater regional groups, and not termed with "Appalachian Transition," institutions in the Pocahontas and adjacent coalfields have arisen in recent years to address economic and human potential there beyond coal, confronting both direct and indirect issues of job loss and automation. Long-standing groups such as Big Creek People in Action in War, West Virginia, the Council for the Southern Mountains, the Appalachian South Folklife Center, Community Connections, and the South Central Educational Development Corporation have been working in this region for years on health, education, senior issues, youth outreach, and so on. More recently, associations or nonprofits have attempted to take on leadership in the arts (Gary Bowling's House of Art, RiffRaff Arts Collective) and business innovation (the MIT FabLab partnership of Pastor Travis Lowe and the Center for Applied Research Technology at Bluefield State University).

Despite technological interventions and advancement, the coal industry did not usually share tech modifications or advances across companies, nor did they support a local culture of invention.[14] Because of the stasis of this local single-sector economy and the conditioning of local workers and management to rely mostly upon this sector and supporting sectors for employment, the kinds of social diversity or entrepreneurial diversity taken for granted in large metropolitan areas, if set up to function in the Pocahontas

Coalfield, can seem and be nothing short of revolutionary. Thus, emerging groups and emerging partnerships can magnify impact, and to do this, need financial and transparent support equal to that of prior coalfield development support.

Clear in all of these partnerships remains a concern with employment, and the very high likelihood of little to no coal-related employment returning: automation remains starkly real and anticipated to continue in most industries left, whether coal, retail, natural gas, or medical.

LOOK TO THE PRESENT AND THE FUTURE, BUT DON'T MUDDLE OR DEIFY THE PAST

What Good Old Days? Whose Good Old Days?

Respondents for *Big Rural* have described that since the region at hand has been in population decline for decades, current youth do not remember "the good old days." They want new models of what their towns or places can be and need reasons to remain.[15] It can be tempting, also, to whitewash the precoal preindustrial history of the region. That the land companies were ruthless in their acquisition of mineral rights exists as a provable fact, and that mining companies exploited and/or controlled miners and other employees is also a well-proven fact.[16]

Since coal offers a complicated legacy on a number of levels (health, wealth distribution, environment), the current transition discourse cannot draw significantly from that legacy or heritage for justification of entitlement to alternate spatial creation.

I do not subscribe to a right of heritage for the region's European colonist progeny. It is terrible to be swindled out of one's legally held land and connived from potential profit from it. Nevertheless, we enter dangerous territory ethically and academically when we assert that the inhabitants in this space prior to industrial coal had a legacy right to be there. Long-time Appalachian scholar Talmage Stanley writes of his family's correspondence from the industrial coalfields and he laments at the treatment of people there prior to coal's industrialization:

> Hidden behind Apperson's mention of the "Poco fields" is a mountain culture that was all but destroyed as the Norfolk and Western Railroad and its subsidiaries sought to wrest control of the mineral and timber rights from the indigenous farmers of Mercer and McDowell counties, in West Virginia and Tazewell and Buchanan counties in Virginia during the last twenty years of the nineteenth century.

Indigenous? No, the farmers there were highly likely to not have been Native Americans (I contend that an ugly streak in the region is the local White claim to a "Cherokee" grandmother—thereby manufacturing a right of occupation or place. As an Affrilachian friend and I joke—both Whites and Blacks in the region like to point to potential Native American ancestry, when we/they likely ought to be pointing at each other/one another); by contrast, their kin likely had a hand in driving those people out by foot, fight, or murder.[17] Thus, the spiral backward for who has a right to be in this region becomes sticky very quickly. That there ought to be fair practices and environmental stewardship does not rely upon an Ur-Appalachian claim to the land. Likewise, the current transition from a coal region to a region having other meaning rests also in a cultural and identity conversion of "coal" families to something else as meaningful. In this, what of African Americans forced to work in the mines or who chose to move into the region? Or of the Jewish, Italian, Hungarian, and Slavic descended people in this space now? They get erased in this Ur-narrative.[18]

That the United States sits upon a history of land conquering and resource theft by actual real people acting collectively becomes absconded in absurd assertions such as mountain White people being or having become "indigenous." Now, the people in these same coalfields are being replaced, purposefully, by machines as the same companies replaced the regional purpose during the wresting of this area into an industrial coalfield. Had these working-class people then developed a right to be there in the coal employment years' interim? What about their managers? The local small boss operators? What about these people Stanley mentions? Do their progeny now have a "right" to be there? To make a heritage- or legacy-based land claim?

Implied but unstated in the "Poco fields" is the rich and diverse culture of African Americans and Southern and Eastern Europeans, who converged by the thousands on the Pocahontas Coalfield beginning in the 1880s.

Stanley references other invisibility in certain artifacts—an erasure of African Americans from this rural industrial narrative. His dissertation reads in its Poco Field section of a long "thick" description of democracy deficit. Though he focuses on the coal baron Koppers, he also illuminates their manipulation of people, the replacement of workers by machines, the fight by workers for fair wages, and the switch over time by these same workers to loyalty to the company rather than loyalty to colleagues.

Thus, the argument of Ur-right and legacy as reasons for rights to land and why certain crimes ought not have been committed against the people on these lands may not get us to the best current conceptualizations of fair or equitable.[19]

REDEFINE REGION, REDEFINE POWER, CONNECT ACROSS THE RURAL

Toward increasing health outcomes, for example, or to increase economic sector participation outside the energy sector, statutory borders may need to be circumvented by other modes of regional organization in order to find solutions or new approaches to dire regional issues.[20] Large organizing interventions may work in single-sector economies but fail when a region's interests do not align with those of the people in power in the individual states.[21] The "Right size" for a social organization may not be fixed or on the binary of macro-scientific rational response or overly determined small-is-beautiful localist solutions.[22] In a subsection of this region, host to a long list of economic, community, and environmental problems, the rural needs to be reconfigured and reassembled, connecting and empowering the people in these adjacent overdeveloped and increasingly ever-more rural sections to link to one another in action and economy, irrespective of statutory borders or power. Science and technology expertise ought to support local people's empowerment. Educational institutions ought to work cohesively and conjointly to solve regional issues of employment, health, environmental degradation, legacies from coal, or other single-sector development. People ought to lobby for transregional approaches to economic development. Training and support for more distributive models of corporate organization ought to be provided. The hyperlocal and marginalized rural there needs to be reassembled into a connected bigger rural through a National Rural Policy or Strategy. Imagine reassembling people and politics in the coalfield subsection in question in the central Appalachian region to work together irrespective of state borders.

Redefine Regional Success

The residents of the Pocahontas Coalfield remain very attached to coal identity and heritage, as most families have had members working at some time in coal.[23] Talmage Stanley points out the scholarship related to imagining a place not attached to economic imperative or to fluctuations in the global marketplace—to a meaning of place beyond capital.[24] I would also state that this would need to be a creation of a region beyond certain state strategic or scientific measurement—two other main reasons for geographic assessment in the United States.[25] But these latter tasks continue even if assessments for global capital's utility remain negative. How does one get an honorable valuation of place now? Who gets to decide? And can scientific, capitalist, state, and conflated configurations of these be usurped or subverted?[26] One *Big Rural* respondent asked this question directly: Why must valuation be tied to capital? From a theoretical postcapitalist standpoint, this would be a great exercise. However, in the meantime, I approach the possibilities for this region with far too much pragmatism for this suggestion to merit high priority consideration. Jobs are not a cure all, but work also matters as there are long statistics for the region correlating good work with good life, good health, and so forth.[27]

Rural people can create an identity not tied to what was the biggest rural (Big Rural) industrial intervention in their area. For example, scholar Rebecca Scott tackles issues of masculine coal identity. Campbell et al. tackle the issues of American rural country identity. In short, a range of work-related identities exists in the rural U.S. context. Coal is not the end all and be all for rural work-related identity in the region in examination.[28]

Search for Other Assets

Social entrepreneurship models can upend the deficit models of community assessment and measurement focused only on what is lacking, and, work instead from assets identified. In the case of Mercer Street in Princeton, West Virginia, the disrepair of Downtown opened an opportunity to recreate Mercer Street as a creative hub, very different in character and personality from its coal and railroad town heyday. [29]

SCIENCE AND TECHNOLOGY FOR OTHER ASSETS: REORIENT ALL STATE INSTITUTIONS OF COMMUNITY AND HIGHER EDUCATION TO SOLVING REGIONAL ISSUES

Imagine how powerful this would be: instead of Virginia Tech's recent focus on autonomous technology, what if it focused instead on achieving in its own very rural backyard, work for all . . . health for all? What if groups of scholars from across disciplines devoted to this their research and inquiry?

RECOGNIZE THE FALLACY OF ECONOMIC DIVERSIFICATION OF THE RURAL IN THE UNITED STATES AS A RESCUE TACTIC

With respect to the popular economic buzz phrase "economic diversification" as a cure for the single-sector rural industrial space, this construct is empty. First, how diverse is diverse? Rarely do its proponents define this; just as rarely, if ever, do they define "economic." With respect to single-sector rural economies, the popular movie and book series *The Hunger*

Games sectioned off the United States accurately: rural spaces have become relegated largely to single sectors. This is not news in the American South, historically a land of tobacco and cotton, now replete with 32 million acres of pine plantations and thousands of contract chicken houses.[30] Or to the mountain South—denuded by the nineteenth century. The environmental sociologist William Freudenberg contended that if a location is not already economically diverse, it will most likely not become it. Said in popular parlance: location, location, location. Metropoles maintain society centers and pivotal access.

Not every rural industrial space has become extractive but many rural spaces in the United States are single sector. With respect to political power, there are more similarities than differences among single-sector areas. For example, similarities in rural industrial single-sector U.S. spaces include:

- Federal and state policies often ignore geographic limitations such as access and do not take hinterlands (spaces beyond the potential to become bedroom communities) into account in terms of the role they play or have played in the local or larger economy, nor take into account what would be best going forward in terms of maintaining community.
- Politicians overpromise the ability to attract in other industries that are not extractive or single sector.
- Politicians keep the focus on bootstrapping or jobs as fixing issues rather than actually fixing issues.
- The isolation of rural spaces from one another keeps the United States from developing a national policy of dealing with areas in the rural depleted in people and other resources. Even if a National Rural Policy is enacted, would lobbyists for powerful agricultural interests in the Midwest and California and energy lobbyists weaken its purported equalizing purpose?
- This disconnect among people across these rural industrial spaces keeps the United States from requiring industry to develop a Plan B for the demise of said industry for the day when operations end.
- Focus on blaming the poor keeps middle class and upper middle class rural people from dealing with/interacting with the people who are left (the poor) when prime jobs no longer are available either due to mechanization, relocation, or resource depletion.
- Seeking to attract in more single-sector industries keeps a focus on an outside source of income as being a panacea.

In the face of these similarities, what if the checks (whatever those are and from whomever) stop? Who suddenly becomes most valuable in the community? How can we then start to value and follow those people now? What would we lose or gain?

STOP EDUCATING THE RURAL TO LEAVE THE RURAL

National educational policy does not consider the kinds of skills necessary to stay rural. Under the Obama Administration, Silicon Valley and larger tech companies had deep access to the White House. At that time, a group of them formed a nonprofit, P21, to promote the kinds of skills they want promoted among students. With the rise of remote work, this skill set may be more applicable also in rural areas than when this group was founded, yet, as a group promoting certain "skills," no small business was represented. If you visited the P21 website, little focus was on the kinds of skills that can help a student mitigate the issues in the rural industrial and stay rural, if working in or with computer technology was not of principal interest. Not that P21 was not useful, it simply focused on many abstract and soft skills that may or may not afford one an opportunity to remain rural.

By way of illustration, the following organizations and individuals were instrumental in founding P21:

Government: U.S. Department of Education

Founding Organizations:

- AOL Time Warner Foundation
- Apple Computer, Inc.
- Cable in the Classroom
- Cisco Systems, Inc.
- Dell Computer Corporation
- Microsoft Corporation
- National Education Association
- SAP

Individuals:

- Ken Kay, President and Cofounder
- Diny Golder-Dardis, Special Advisor and Cofounder

An addendum to this may be, that after a pandemic and recent crisis of work and remote work, people in rural areas may be more competitive for remote work in these soft skills as they can be paid less than workers in more expensive locations. However, the main economic drivers in rural spaces remain natural resource ties. At issue is that large tech companies often now set the overall learning objectives for U.S. students, regardless of whether those skills truly translate across all sectors. For example, in the Obama Administration, his main business advisors and visitors were all from large companies (see the companion figure on the website, www.bigrural.space).[31]:

Education and business priorities from the federal level down remain skewed for large corporations, in other words, promoting priorities nationally for preparing students for work that largely does not exist in rural industrial areas.

CREATE A REAL NATIONAL RURAL STRATEGY/ POLICY BEYOND THE FARM BILL

In *The Failure of National Rural Policy*, in addition to a critique of U.S.' federal policy approaches to the rural, William P. Browne chronicles the development of science-based and driven agriculture in the United States in the nineteenth century and its massive acceptance by universities and by farmers in the early twentieth century. His framing of the introduction of science into U.S. agriculture unsurprisingly follows similar trajectories in U.S. industrial development during this same time: a period of industrial, academic, political, and scientific specialization as well as social and geographic upheaval. In short, most science in agriculture did not focus on mountain or forest farming and the Midwest and West came to dominate rural and agricultural policy and decisions in U.S. federal policy regarding the rural.[32]

Rural scholars Tolbert, Irwin, Lyson, and Nucci in "Civil Society, Civic Communities, and Rural Development" hammer home that rural communities are part of the global economy, and that large corporations set the agenda there. Rural folks can be blind-sided, even when they do everything "right":

> Communities that did not invest in their infrastructure or in their residents, as well as those that did all the right things but were unable to attract business, were simply "forgotten."[33]

Rural America did not neatly follow the plan of free market capitalism. Regeneration and rebirth in a Schumpeterian model do not arise in places with multigenerational poverty, uneven wage attainment, and splotchy economic recovery . Rural became "Big Rural" as even the largest family farms became owned by national and multinational corporations, likewise with mining companies, forestry, and paper.[34]

The current U.S. rural approach promises more of the same, looking to big corporations as guides for what kinds of education and economic policies should prevail. As Big Corp further subsumed Big Rural, the potential for a national level or local government intervention with respect to setting an agenda for rural development has become increasingly weak, as evidenced by big corps' "sucking in" of rural industries in the 1970s and 1980s.[35]

The nation-state is not as useful (but still somewhat useful) in conceptualizing the movement of capital currently at stake. Though trade was international, local and national policies still obtained. With transnational corporations, NAFTA, IMF, World Trade Organizations, how much of local is controlled locally? What does local mean? How local is local? In short, in the rural industrial space, how bad does it have to get before inhabitants advocate for something else? Are they too hamstringed to reassess, too threatened by a living death? Is there a bottom to hit or is it a bottomless pit?[36]

NOTES

1. Daniel Breslau, "Economics Invents the Economy: Mathematics, Statistics, and Models in the Work of Irving Fisher and Wesley Mitchell," *Theory and Society* 32, no. 3 (2003): 379–411; Deborah Stone, *Policy Paradox: The Art of Political Decision Making* (New York: Norton, 2001).
2. Stone, *Policy Paradox.*
3. Ibid, 128–30.
4. Wiebe Bijker, Trevor Pinch, and Thomas Hughes, *The Social Construction of Technological Systems: New Directions in the Sociology and History of Technology.* (Cambridge, MA: M.I.T. Press, 1987).
5. For example, here is an overview of federal support for coal-related research: https://www.nap.edu/read/11977/chapter/13 at the National Academies Press.
6. James C. Scott, *Seeing Like A State: How Certain Schemes to Improve the Human Condition Have Failed* (New Haven, CT: Yale University Press, 1998), Ch. 10.
7. Jennifer Stoll, Jessie P. H. Poon, and Trina Hamilton, "Sustainable Practice? An Examination of Canada's Agricultural and Energy Cooperatives," *Professional Geographer* 67, no. 2 (2015): 187–94; Joyce Rothschild, "Workers' Cooperatives and Social Enterprise: A Forgotten Route to Social Equity and Democracy," *American Behavioral Scientist* 52, no. 7 (2009): 1023–41; Sammy Zahran, Samuel D. Brody, Arnold Vedlitz, Michael G. Lacy, and Chelsea Lynn Schelly, "Greening Local Energy: Explaining the Geographic Distribution of Household Solar Energy Use in the United States," *Journal of the American Planning Association* 74, no. 4 (2008): 419–34.
8. Stanley, "The Poco Field: Politics, Culture, and Place in Contemporary Appalachia," (PhD diss., Emory University, 1996), 65.
9. Edna F. Einsiedel, Erling Jelsøe, and Thomas Breck, "Publics at the Technology Table: The Consensus Conference in Denmark, Canada, and Australia," *Public Understanding of Science* 10, no. 1 (2001): 83–89.

10. This may be a farfetched assertion, but we currently understand the present state of science for hire and what this has done to knowledge production.

11. The Appalachian Regional Commission, for example, supports no academic research.

12. See: L. Spencer, "Decades will be needed to fully assess fracking's impact on drinking water, Geological Survey says," *The Rural Blog*, University of Kentucky, Institute for Rural Journalism and Community Issues, June 1, 2017, accessed July 21, 2023, https://irjci.blogspot.com/search?q=decades+will+be+needed+to+fully+assess. For the underlying study, see: U.S. Geological Survey, "Methane and Benzene in Drinking-Water Wells Overlying the Eagle Ford, Fayetteville, and Haynesville Shale Hydrocarbon Production Areas" by Peter B. McMahon, Jeannie R.B. Barlow, Mark A. Engle, Kenneth Belitz, Patricia B. Ging, Andrew G. Hunt, Bryant C. Jurgens, Yousif K. Kharaka, Roland W. Tollett, and Timothy M. Kresse, *Environmental Science and Technology* 51, no. 12 (May 31, 2017): 6727–34, https://doi.org/10.1021/acs.est.7b00746.

13. Ken Ward, Jr., *Coal Tattoo* (blog), http://blogs.wvgazettemail.com/coaltattoo/

14. Dan Hodge, "Appalachian Coal Industry, Power Generation, and Supply Chain," By Hodge Economic Consulting, *Appalachian Regional Commission* (March 16, 2016), https://www.arc.gov/report/www-arc-gov-power/.

15. Interview with qualitative respondent, 2014.

16. John Gaventa, *Power and Powerlessness: Quiescence and Rebellion in an Appalachian Valley* (Chicago: University of Illinois Press, 1982); Stanley, *Poco Field*; Jerry Bruce Thomas, *An Appalachian Reawakening: West Virginia and the Perils of the New Machine Age*, 1945–1972. Vol. 12 (Morgantown: West Virginia University Press, 2010).

17. Howard Zinn, *A People's History of the United States* (New York: Harper Perennial Modern Classics, 2005).

18. Ronald L Lewis, *Black Coal Miners in America: Race, Class, and Community Conflict*, 1780–1980 (Lexington: University Press of Kentucky, 1987); Stanley, *Poco Field*, 69.

19. Rebecca Scott, *Removing Mountains: Extracting Nature and Identity in the Appalachian Coalfields*. Quadrant (Minneapolis: University of Minnesota Press, 2010).

20. Respondent, 2015. For example, the region's opioid use and economy crosses regional statutory borders, but the tactics for curbing its impact and for treatment stop mostly at state borders.

21. I draw these conclusions based on interviews with grasstops and other stakeholders in the case-study region and also by following funding to those regions to address dire needs.

22. There is no one-size-fits-all solution and even the charms of the localist movement can fall apart when locals reveal prejudice, lock out others due to provincialism, fail to bring people at the margins to the table, or engage in cronyism and corruption. There is a long literature on the latter in extractive Appalachia and an entire literature of whether such a thing as a "resource curse" exists in terms of extractive areas diminishing democratic potential.

23. Rebecca Scott, *Removing Mountains*.

24. Stanley, *Poco Field*, 45. He draws on Herbert G. Reid, "Global Adjustments, Throwaway Regions, Appalachian Studies: Resituating The Kentucky Cycle on the Postmodern Frontier," unpublished paper, October 1995, held by the author.

25. David Harvey, *A Brief History of Neoliberalism* (New York: Oxford University Press, 2007); James C. Scott, *Seeing Like a State*, 1998.

26. Travis Lowe, "I've Seen The Future. It Looks Like Appalachia: Life after the Job Apocalypse," *Medium*, March 31, 2017, https://medium.com/startup-grind/ive-been-to-the-future-it-looks-like-appalachia-5553bf89f275

27. Shannon M. Wooley, Susan L. Meacham, Lauren C. Balmert, Evelyn O. Talbott, and Jeanine M. Buchanich, "Comparison of Mortality Disparities in Central Appalachian Coal- and Non-coal-mining Counties," *Journal of Occupational and Environmental Health* 57, no. 6 (June 2015): 687–94.

28. Rebecca Scott, *Removing Mountains*; Hugh Campbell, Hugh, Michael Mayerfeld Bell, and Margaret Finney, eds. *Country Boys*: *Masculinity and Rural Life* (Penn State University Press, 2006).

29. Richard Valencia, *The Evolution of Deficit Thinking*: *Educational Thought and Practice*. *Stanford Series on Education & Public Policy* (New York: Routledge, 1997).

30. Thomas R. Fox, Eric J. Jokela, H. Lee Allen, "The evolution of pine plantation silviculture in the Southern United States," in Gen. Tech. Rep. SRS 75, Asheville, NC: U.S. Department of Agriculture, Forest Service, Southern Research Station (2004), Chapter 8: 63–82. https://www.srs.fs.usda.gov/pubs/9647; Suzi Parker, "How poultry producers are ravaging the rural South," *Grist Magazine* (February 22, 2006), http://grist.org/article/parker1/

31. Jeffrey R. Brown and Jiekun Huang, "All the President's Friends: Political Access and Firm Value" (NBER Working Paper No.w23356, April 2017). https://ssrn.com/abstract=2961081

32. William P. Browne, *The Failure of National Rural Policy* (Washington, DC: Georgetown University Press, 2001).

33. Charles M. Tolbert, Michael D. Irwin, Thomas A. Lyson, and Alfred R. Nucci, "Civic Community in Small-Town America: How Civic Welfare Is Influenced by Local Capitalism and Civic Engagement." *Rural Sociology* 67, no. 1 (March 1, 2002): 90–113, 229.

34. Ibid, 229.

35. Ibid, 230.

36. Alessandro Bonano and Douglas Constance, *Stories of Globalization*: *Transnational Corporations*, *Resistance*, *and the State* (Penn State University Press, 2008).

Chapter 4

So, What of a National Rural Strategy?

The journey to understanding the invention of the Pocahontas Coalfield and its most recent manifestation as a place of deep struggle—environmental degradation, depopulation, lagging seventy + years behind in income achievement, comorbidities, addictions—landed me in a place of examination usually out of consideration for most social change agents: policy. For many of us, the importance or even the relevance of policy and how it is set escapes our everyday reality. The news, newspapers, and so on feature politicians, not policies.

We are exposed to political spectacle but not to policy practice. Even if we have been lucky enough to learn how a bill becomes a law, we lack exposure to the import of the broader strokes of setting direction through appropriations, requests for proposals or applications by and to federal agencies, and the divvying up of money to states. These broader strokes center on policy. This policy comes to us not by the magical thinking of politicians, but often through partisan think tanks or large corporations promoting an approach, and successfully achieving policy strategy or law.

On the ground, we drill down and often focus on the bleeding person in front of us. In the U.S. context, the immediacy of people in need and without a safety net or community support not only stands as obvious and beyond being moralized, but becomes even more urgent by our system of competing for resources in an artificially made competition invoking scarcity where there does not have to be any. Said differently, the social care realm in the United States addresses the needs of people and communities when the government and private industry choose not to step in. This is set up unnecessarily as a competition where there will never be enough resources to address the needs of people and places for achieving basic services and education—the hallmarks of modern dignity and respect under the State.

Nonprofits compete for government-allocated and private foundation funding. Many places in need lack the people or the institutions to enter this competition, and, when they try, they lack the resources that long-standing nonprofit actors maintain both in connections and staff capacity. The places where people have the least resources, like the Pocahontas Coalfield now, and the fewest institutions that can apply for additional government or foundation assistance, remain without additional assistance beyond individualized assistance in the form of benefits to individuals such as direct public assistance. In places competitive for these "extra" funds, individuals do not have to rely only on their own relief received directly: nonprofits may take up the slack by providing lower cost services such as healthcare, counseling, tutoring, and so forth. Larger municipalities subsidize affordable public transportation, healthcare, emergency services, senior services, and so forth. They receive more state funding, and, with proof of serving larger populations, are more competitive in the applications for additional funds.

How else could this be organized other than how it is today?

How else ought it be organized?

Imagine if a federal agency like the Appalachian Regional Commission (ARC) were funded to the degree of the Department of Defense or Department of Energy, or, by contrast, took its allocation for four or ten years and concentrated it in one place to address all the issues in one place, then moved on to the next?

What if the U.S. government focused on fixing the systems that cause such economic disparity among individuals and places?

What if we taxed the rich and corporations and disallowed them from stashing their funds into foundations that serve as tax shelters and which are controlled by a few individuals deciding from a distance what communities and places deserve assistance and which ones do not? What if instead that stashed money were available through taxes where ostensibly we citizens could vote for representatives to decide what happens with this money?

In select other countries, the government assumes the responsibility for much of the work completed in the United States through nonprofits, religious charities, and foundation gifts. The allocation of resources gets set through government policy designed by elected representatives or those they hire to provide input and research.

In the United States, policy also gets influenced by lobbyists, corporate giving, and in some cases, also graft. In *Big Rural* I cover corruption as "tradition" in twentieth-century politics in West Virginia, highlighting the research and first-person experiences of Huey Perry in Mingo County and investigative journalist Thomas Stafford and his critique of the political dynasty of the Moore family. On top of this, we have this relief system outlined above of addressing individuals affected by economic or other shifts, rather than addressing the systems contributing to the need for relief.

Places with resources even may have a hand in promoting one federal funding priority over another. States will hire experts to assist with research to support one favored direction over another.

The short lesson in all of this? Places with tax resources have resources to hire people to do research as well as lobbyists to advocate for funding and to influence politicians that set policy, and these places with tax resources remain in that cycle. Places without access to these tax resources remain perpetually unable to compete, and if unable to compete, relegated to being unworthy of funding, undeserving of tax allocations set through policy by people hoarding these tax allocations elsewhere, or unable to compete for charitable giving doled out based on the ability to compete in the race for money designated by foundation employees and the foundation's wealthy trustees rather than that money being assessed as taxes and then decided upon by elected government actors or at least by government hired staff (the Fifth Estate—non political appointee people in government employ).

Without a doubt, cracking the façade of this system and questioning whether this is what ought to be done comes with risk. In the Appalachian Region, those doing on-the-ground work can't dare question the priorities of an agency like the Appalachian Regional Commission (ARC) without real risk of being cut out of the competition for already scarce funds. The ARC also lacks transparency in its processes, as does USDA Rural Business Development. Unlike in science or scholarship where peers decide the merit of your work, in the case of these federal agencies, nonprofits submit grant proposals during a short window open for competition. These proposals are not scored by peers, but by anonymous reviewers, whose expertise for judging merit is not open for review by the nonprofits in any participatory fashion or open to any public scrutiny. As funds come in bulk from the federal agencies to state agencies, where Colorado may host an open competition for surplus funds such as recent ones through the Biden Administration's Build Back Better initiative, West Virginia may not open any competition, doling out and assigning funds without any public process or accountability.

Much to discourage engagement in on-the-ground social change work flows unabated. Collectively working in Southern West Virginia, pursuing change work, groups I have worked with or interviewed have been discouraged or dismissed by state agencies, local city officials, school principals, city managers, county commissioners, federal agencies, powerful regional foundations, regional social capital nonprofits, banks, lenders, community lenders, community foundations, mayors, businesspeople, hired small business developers, agricultural or veteran focused nonprofits, economic development professionals, economic development agents, economic development departments at R-1s, Extension agents, University Extension, health equity-focused foundations, and local and state politicians.

Big Rural points to cracks of light in the canopy toward starting and directing engagement even in McDowell County, WV, the county declared the most hopeless place to live in America. Little wonder that respondents, when I interviewed grasstops and others in and around the Pocahontas Coalfield, declared lack of vision and leadership as the top issue. Some simple questions could be—how can anyone stand by and let this happen, let a place become "hopeless," and once that has happened, how can a person continue to stand by?

ODE TO ENCHANTED LIGHT

Even in the immense obstacles posed by this system, light shines through cracks in the canopy. Any approach I take in or discuss in this volume, that respondents take, that the oral history participants take, also stands open for other people to attempt, and in the volume's conclusion, I list and reflect on these.

Big Rural stands also as an act of self- and regional-determination.

We can get to others and discuss our assumptions, we can decide our ethics, and rather than only complain or critique the system, we can put into that void *what ought to be*.

Because neither the Cavalry nor Calvary is coming. If we don't do it, someone else will, and they already are by deciding that who we are and where we are is not worth the money, even the money that we pay or someone else earned off where we live. Or, if they decide something we live is of interest, they will cut us out and decide how to proceed as Huey Perry chronicles.

Here I am looking directly in the face of the academic fear of declaring what ought to be as a form of teleological historicism, that is, that history follows laws. I am proposing the opposite—that although I declare what ought to be, I do not state that what I propose is immutable, or, follows as natural or more natural. I am aware that what I propose is contingent.

What follows in this chapter and the next are not only thought exercises. For the National Sustainable Agricultural Policy, I also apply the lessons learned from working as a sustainable farmer (in addition to the other hats I wear, my husband and I have a smallholding farm).

In a democracy, all policies ought to make people more free.

Citizens can also propose policy, and, also ought to participate in policy in an active democracy. We ought not to offload this onto politicians, wonks, or bureaucrats much less onto lobbyists.

As a researcher, as an academic, as a person engaged in social change, but most importantly, as an informed citizen, in order to explore what a national rural policy might enable to make a rural industrial space such as the Pocahontas Coalfield more democratic, I set out here an additional range of issues to keep in sight, but, for the sake of space and efficiency, cannot fully elucidate without veering too far from analyzing Big Rural as a power dynamic. As an organizing force, Big Rural crosses spaces and shares characteristics with the developments of mega-industries rooted in the onslaught of European global colonialism. This force shifted in character but did not stop with the end of that colonial era. Any number of modern phenomena from white collar work to supermarkets to television to trains and trucking to shipping and logistics functions to mask the mechanics and varied costs (economic, social, environmental, community, personal) of Big Rural's maintenance. Appalachia as a kind of internal U.S. colony has been well-debated in other scholarship; in this respect, Appalachia is not unique but one of a type of Big Rural, a type with a set of circumstances in common with other rural spaces dominated locally but also far flung globally with throughlines and connections through single sector dominance.

Many issues contained in the rhetorics of the rural in the United States can make the task of compiling a general national U.S. rural policy appear futile, for example—popular culture, political, and academic fights over who speaks for the rural, how the rural may stand in for other issues, and all the rhetorics of associations of the rural space: American progress, jobs, environmentalism, big Ag, the localist movement, local Ag, idealizing the White male rural, idealizing the current state of nature; issues of gender, race, class; deficit models, degrowth, spatial realities of services, myths of self-reliance, myths of government largesse, etc. Nevertheless, policies, or another federal "policy" document, the strategy, would be a welcome step in addressing how we as a nation ought to assess, understand, and approach the myriad of issues in a space like the Pocahontas Coalfield, especially with respect to ensuring and safeguarding democracy in these spaces.

In large strokes, the focus in *Big Rural* on the Pocahontas Coalfield brings to the fore major issues in the rural industrial space that scientists and engineers assist corporations and government to enable. What if a similar strategy instead focused on economic sectors in rural spaces in which jobs are less likely to be automated away? And R&D also focused deeply on solving issues of rural health? Rural sustainable business? Rural land, plant, and animal environments? The list goes on.

The issues in the Pocahontas Coalfield are not unique, and are experienced in many rural places across the United States, and, the globe. In an ideal situation, the United States would not continue to subsume national rural policy under agricultural policy (including economic policy) under the Farm Bill.

We would join nations such as Canada, Chile, Finland, France, and Korea with explicit national rural policies or plans.

In the absence of such explicit national strategy regarding rural spaces, nothing prevents me from outlining potential policies to be worked on jointly by stakeholders such as nonprofits, local governments, and educational institutions. If individual states or legislators or legislatures choose not to engage, then other regional grasstops (those who organize the grassroots or are leaders in those entities, often with an overview of what systemic shifts need to happen, or, at least, a sense of what policies are needed) or faith-based constituents could move forward with cross-border and extra-statutory connections and coordination. In short, if the government (little or big G) cannot or will not address issues in rural subsections, people in those subsections ought to join together for mutual benefit and affinity.

If states are not addressing the issues in the rural, then regional grasstops ought to commit to working in subsections to make their work more effective. The federal government ought to continue its support of cross-border and extra-statutory work in the realm of rural economic development regarding the need to connect adjacent subsections with more financial and earning experience in emerging economic sectors.

In addition to forming regional policy and coordination on issues of health, declining population and smart shrinkage (focusing on the quality of life of a place in population decline and working to make the remaining people stay and the environment and community sustainable long-term), and explicit discussion of regional vision, economic sectors, and the kinds of research and development to be supported in the region, regions must connect to other similar rural regions and work together. By way of example, it is time for central Appalachia and other rural spaces to stretch beyond nineteenth and twentieth-century work-affiliated identity out to other economic sectors, linking to and building on an agricultural and pastoral past and embracing emerging sectors and long-standing cultural strengths for a creative, neighborly, and hospitable future.

What follows is a snapshot potential strategy with a meta-analysis of components of a national rural strategy that could involve research and development, scientific and technical expertise, education, economic policy, and democratic participation. I am not drafting a full national rural strategy. I also note that since I started researching and writing this volume over a decade ago, the idea for some kind of national rural policy has found favor in opinions for the *New York Times* and advocates in think tanks. While I have, with consideration, orchestrated the present volume to be rooted in research, I also open it to the scholarship of thought experiment, imagination, practice, potential, and new power relationships to policy.

Let me state again: if the government, locally or federally, will not advocate for policies and practices needed in a rural space, then grasstops can, as they are able, organize policies and practices themselves. I add that caveat of "as they are able" because in some circumstances, staying alive in the face of single sector domination imperatives may be all that rural locals can do, physically, emotionally, and socially.

I am also calling here to pull the concepts of rural *in the rural* up from eye-level practice to rural as having an organizing force. I could write an entire second volume on only a handful of the shortfalls of the local foods movement and sustainable agricultural movement (if the latter is even a movement) in terms of the organizing forces of modern life. The best analogy I can muster is that hyperlocal foods advocates and producers have much akin in philosophy with the Doomsday Prepper movement (preparing one's family to endure longer-term shortages of currently readily available commodities—food, and energy being the top two): source what you need in terms of technology—tractors, mechanisms, tools, seed—from beyond your community in order to fortify your own or the needs of a scattered few. Conceptually what links this personal or hyperlocal focus stands as the lack of regard or consideration of larger organizing forces, such as economic sectors and all the stakeholders, research, resource analysis, and political force that goes into organizing and maintaining those. Little to no federal funding mechanisms are really addressing organizing regional or local food production as sectors, with all the needed infrastructure underwritten in federal, state, and private partnerships akin to opening a coalfield in the nineteenth century. The dollar investment for that shift would be massive, well beyond that currently in play for renewable energy, as food production organized and dispersed regionally would be far more decentralized than is current municipally-focused energy production with its physical limitations due to the mechanics of the energy grid and energy distribution.

Before I step into the even wider water of broad stroke policy, let me state all of this even more plainly, or with an even better analogy, if I can.

Although some 50,000 homes may get their energy from a single source, in each of those homes, we all are eating very different things, sourced, at a minimum, from places as disparate as our own backyards to industrial plantation productions 5,000 miles away. In most cases, we don't know (or care) where we get our food or energy from, we just care that they are available. To have easy, seamless energy, we will put up with being told where and how we must source it and how much we will need to pay for it. In this respect, while dominating what is possible in our modern lives, seamless energy, for most folks, carries no status, no cultural associations, no marker of independence, no identity signifiers (even those folks in coal country don't really know if coal produces the energy at their own homes), no memories, no visceral reactions, no allergies, no aversions, no obsessions, no celebrity, no shame, no sensuality, no adornment, no attention.

By contrast, despite easy access to food in Western society, none of us wants to be told what to eat. Those of us who grow it do not want a Soviet system where we are told what to grow.

Yet, that ease of access at the supermarket, that casual filling of our shelves with novelty food from 5,000 miles away, means someone somewhere has made a choice for sufficient surplus production. We participate in a system that masks the farmers and farmworkers who do that work. We participate in a system that masks the physical, economic, community, personal, and psychological issues of these farmers and farmworkers, that hides their labor from view, and, from discussion. If we truly reorganized food production even in regions akin to those the size necessary to, or delimited by energy production, many of us would come face to face with the struggles of farmers and farm laborers (healthcare, housing, pesticide use and exposure, mental health, etc.) that currently get rendered seamless and invisible in Big Rural:

That agricultural production in the Big Rural has been organized through infrastructure such as processing plants, warehouses, roadways, trucking, railways, shipping, freight, and so on into an economic sector. That original resource-based production has been identified as having the potential to scale, and, that that assessment was conducted by scientists and engineers. That that assessment made the case for someone to achieve state-sponsored and/or private, or corporate financing to get that production happening and scaled.

Currently, corporate commodity agricultural productions are subject to global forces with massive infrastructure. The local foods movement in the United States focuses on individualized local producers, usually with little to no focus on infrastructure above the level of a small food aggregation center, a shared refrigerated truck, a commercial kitchen, and occasionally a local poultry processor or livestock kill floor.

Something is possible other than global forces, the hyperlocal, or super-individualized structures of organization. None of these models addresses the practical long-term needs of our changing climate, decimated soil and environments, or growing population. The latter (the hyperlocal and super-individualized) ignores how seamlessly our *current* agricultural infrastructure satisfies not only calorie and nutritional needs but a whole host of other social, emotional, and practical imperatives enabled through infrastructure, science, and funded and supported through policy.

Envisioning other kinds of economic sector organizations must be part of any serious discussion of rural industry.

And the government is not going to bring you that vision or grant you permission to break beyond what it calls for in its current lineup of requests for proposals.

You can identify local problems. You can conceive of new solutions.

This series of thought exercises in this section of Big Rural stands to grant you any permission you need not just to envision or dream, but to analyze and act.

Can we break through these models in agriculture, or their equivalents, in other rural productions? Conceptually? Practically? In policy?

This thought experiment in a national rural strategy serves as a loose template drawing on the elements that enabled the Pocahontas Coalfield—science, technology, research, finance, and government financial support. I do not rehash the issues explicated in *chapter* 1, and I do not review statistics, positive or negative, reflecting the quality of life and robustness of democracy in the rural United States—though these would be obvious parts of an actual national rural strategy.

Let me also state that I am under no delusions of expecting the following strategy to be adopted by a federal agency. The best and most likely outcome of any of what I propose below is that something from it gets incorporated, uncredited, into nonprofit or state policy, as, in the social change world, wheels are rarely invented, just copied or retooled or borrowed. This is not my first system analysis rodeo, but here this particular one goes out of the gate, to hold onto as long as it makes sense.

NATIONAL RURAL STRATEGY

In the spirit of the United States National Security Strategy, updated usually with each entering federal administration, the following National Rural Strategy addresses a range of issues explicitly affecting sections of the United States with fewer than 50,000 inhabitants. For this, I continue to rely on what had been a definition standard of the USDA, though as of this publication, the Census has moved toward a redesignation of rural to housing units.[1] Rather than approach issues in these sections through the Farm Bill, as is largely currently accomplished with respect to the federal government's approach to rural spaces, this National Rural Strategy stands on its own. For the sake of brevity, I also do not outline the kinds of rural places nor redefine rural industrial.

Coordinate Federal and Intrastate R&D for Additional Economic Sectors beyond Single Sectors and Support Work and Jobs that Cannot Be Automated Away.[2]

Large-scale industries seek efficiency and to eliminate labor by humans. Humans unionize, get sick, demand health insurance, need retirement payments, take time off, need vacation, can produce errors, and so forth. This directive remains as true in clean energy as in coal or other energy sectors, or in large manufacturing. Clean energy is important for many urgent environmental reasons but does not offer a long-term jobs solution. Emerging economic sectors in rural areas must focus on high-touch sectors in which human labor and management remain relevant into the future.[3]

Places that have not yet been developed by multiple industries or by a single sector, or even a single operation, ought to require entering industry to create a sunset fund to set money aside for relocation and retraining of workers and their families once they are no longer needed either due to technological advancement or the end of operations at that site. This is separate from the severance taxes on industries such as coal, which states can choose to have allocated or dispersed beyond the immediately affected workers or location. Countries such as Sweden and Japan have developed sunset fund requirements for factories or operations located in previously nonindustrial rural areas, with the understanding that no operation lasts forever and that the burden of retraining and relocating ought not be externalized onto the state.[4] Furthermore, the workers who relocated to work in that factory or industry ought not to be responsible for bearing the burden of relocating after the factory or industrial site closes or downsizes (as is expected that it will). All the accompanying businesses, schools, and institutional and real infrastructure built to serve the needs of people associated either directly or through locale with that industry must also have plans in place for the eventual downsizing due to technological or industry shifts.[5]

Understand the Resource Wealth and Follow Their Markets. Develop Economic Policies and Procedures Irrespective of Supposed Technological "Advance."

Despite energy company bankruptcies and asset shifts, as the central Appalachian and adjacent region remains resource wealthy, companies are unlikely to shed their land assets.[6] Thus, unlike urban depopulating areas with land opening up for complete repurposing in and by a different economic sector (think brownfield into a museum or housing development into an urban farm), lands currently owned by resource companies mainly will remain owned by resource companies (think coal to timber, coal to natural gas, coal to water, maybe coal to rare earth elements—still resource driven). Natural gas development remains large, with the potential to access the Marcellus Shale through coal-mined land real and high.[7]

Anticipation of future use due to the development of technologies that provide access to, or demand for, mineral or other resource wealth remains a standard expectation in the energy sector. New technologies give access to below-the-surface minerals previously inaccessible; new processes allow for cost effective refining of materials once either too cheap to harvest (no profit incentive) or too costly (can be gotten elsewhere). This technological determinist stance anticipates technology will always evolve to reactivate "fallow" resources.[8]

Likewise, the U.S.' cultural anticipation of technology always producing positive economic impact and new economic sectors arising fully to replace antiquated or shifting sectors does not bear out in workforce, environmental, or population statistics, especially in rural industrial spaces such as those in central Appalachia.

Unlike in urban areas, in the rural industrial coal producing regions proper, the emerging economic sectors (agricultural, liquor, tourist, creative community sectors, etc.) must contend with limited access to land and buildings for their sectors. Land not in use or allocated for conservation ought to be up for consideration for local citizen use for sustainable enterprise..[9]

Worth stating again is that larger scale industries focus on technological advancements that eliminate workers.[10] For the remaining people living in the region, work promoted and supported ought to be that remains tenable into the future and which could steward future generations.

Support Transparency and Technology for Democracy to Increase Democratic Participation.

Corruption and democratic power imbalance in practice remains an issue where local residents mainly work in one industry. Totalitarian work/living space and/or low social trust erodes the capacity for democracy to function in a rural industrial space where one economic sector or employer dominates. Means and modes of transparency become especially important in these scenarios.

The U.S. devotes much attention to increasing democracy in other countries and ensuring transparency. In its own rural "backyard" the United States must implement programs of leadership, technology, and investigative citizen journalism to add more safeguards to American democracy in its rural contexts.

Some technology that supports these aspects of democracy has also made a radical shift in the last fifteen years and an even more radical shift in the last two to three years. This technology, mobile phones, mobile web, and the internet (shortened here to MMI), given a democratic regulatory environment, is people-powered. It is a technology whose content, in a democratic regulatory environment, can be utilized by nonmedia or technology professionals, or by nonstate-sanctioned commentators, to check and balance power. Training in this technology and access to this technology ought to be part of a rural broadband policy. Rural broadband is not a cure-all for rural poverty. Broadband is only as useful as people have access to technology to make the best use of high-speed internet, including its use for democratic as well as commerce and information seeking purposes.

Groups like the Central Appalachian Network and the Appalachian Funders Network provide essential grasstops cross-border leadership in the region. Their Appalachian Transition Fellowship (now defunct) once provided one entry point into grasstops leadership by stewarding capable beginning career candidates into social change work.[11] Yet, more must be done. However, an extra-statutory (cross-border) approach and coordination of civic leadership or participation are essential.

The low democratic participation myth of extractive communities is indeed a myth, as Norway is a very extractive-dependent country and is very democratic with high levels of voting and civic participation, and low levels of corruption.[12]

Until coal became supplanted by oil and gas as a U.S. energy and economic driver (by World War II), workers in the coalfields actively worked for democracy by advocating for unions, shorter workdays and benefits.[13]

Although efforts at democratization in central Appalachia were historically corrupted in the post–United Mine Workers of America (UMWA) formation era, this does not mean they ought not to be tried again.[14] Social media and tracking online corruption through on-the-scene video documentation change the scenario for corruption mapping or reportage.

More people must be trained in how to assess and to address economic, political, educational, and social change. The kinds of programs the United States supports abroad through USAID to promote democratic reform, train local citizens in democracy, and to promote youth leadership ought to be funded in rural America.[15]

Break and Conceptualize Regions beyond Sector Flows and Regional Economic Sector Flows beyond Single Sectors and Urban Commutersheds.

Despite black-boxing and legitimizing state borders in the United States context, in fact, in terms of the flow of people and flow of commerce, states rarely best define either. For example, a region can be broken into primary historical economic interests, and, these specific economic interests often have superseded local and state governmental power as regional power structures (think the power of a king, then reflect on the analogy and power of King Coal or Big Ag or the Tennessee Valley Authority).

Another view of examining how regions relate includes examining the regions through media market saturation. By way of example, the nonprofit-business sector collaboration in the Pittsburgh metro region, the Power of 32, defines its service and economic area by media market.[16] The gas industry in the United States by shale plays. The coal industry by coterminous coalfields.[17] In a regional subsection, coalfields or shale plays or media markets often define cultural, interpersonal, and business relationships. To this, add in proximity to major metropolitan areas, and, according to economic and urban geography literature, most areas within an hour of a commute from a city (within a commutershed) usually find much of their economy tied to production or in relationship with their closest urban center.[18] Political allegiances and alliances often follow these geographical-industrial or business sector orientations.

Political influence has often followed the money and how industries cross state borders. A region can be broken down along economic sectors or main industries nearly as much as it can along state lines.

Large-scale extra-statutory (across state borders) coordination and support for emerging economic sectors ought to be provided, including significant investment in federally sponsored R&D. The federal government and non-profit sector ought to provide concrete meetings, support, and policies for promoting cross-border and extra-statutory approaches in economic and community development, with incentives to link to more prosperous adjacent regions.[19]

Brownfields Research and Rehabilitation for Other Economic Sectors.

The research for brownfield rehabilitation for agricultural and recreational use in rural industrial areas is not coordinated, and, according to rehabilitation scientists, what is known often is not sufficient to guarantee public safety, for example, of consumable products.[20] More resource support for brownfields research and rehabilitation to support the emerging sustainable economic sectors in which jobs cannot be automated away must be provided.

For example, it is not easy to find the correct combination of expertise with respect to anticipating issues of agriculture for human consumption on former surface mining sites, and, especially, on sites near current or past deep mines in the central Appalachian coalfields. Often each site is different and requires different protocols. What works at one site and is true of one site, may not be true or work at another site. Science and ongoing assessment on each proposed site are key to allowing local citizens to understand the economic limitations and potential of their brownfield sites. Scientists ought to offer this to these sites rather than wait for people from the sites to seek them out.

Support More Scholarship on the Rural—Medical, Public Health, Epidemiology—and Comparative Scholarship.

As demonstrated in earlier sections of this volume, major scholarship focused on rural areas of the United States includes rural studies, rural sociology, geology, various regional studies, agricultural, forestry, reclamation, and scientific and engineering energy sector scholarship. In recent years, scholarship on deindustrialization has focused largely on urban areas.[21] Urban studies and urban planning as fields include scholarship on many of the issues also facing rural industrial spaces: depopulation, poverty, brownfields, economic revival and renewal, and crime . However, rural studies as a robust field in the United States faded from universities largely by the 1960s.[22] Most studies on addressing addiction, brownfield, or work-related exposures focus on urban dwellers, with exceptions being very recent work focused on rural areas in public health, epidemiology, and so on.[23] Robust and committed cross and interdisciplinary work is needed again in the rural United States, which brings together industry, economic, health, and other scholarship into dialogue toward developing programs of research and development for rural planning and stewardship.

Fund Innovation and Entrepreneurship and Community-Based Technology Initiatives in the Rural.

Rural Americans take much personal risk with respect to physically demanding labor and personal consumption habits.[24] Boom and bust single sector economies promote dependence upon a single sector for employment.[25] Often appropriate emerging economic sector risk is not well-supported by state, federal, or private foundation or bank funding or by educational institutions.

Rural Americans may know best what kinds of technological inventions suit their local environments and economies. The federal government ought to support the creation of Community-Based Innovation Hubs and Patent and Invention Stations. These spaces would bring together diverse and divergent elements, such as:

- People with heritage knowledge and skills (think: know how to do the basics of community using the basics)
- People with modern (last 100 years) knowledge and skills (think: mechanics, fabrication, welding, electrical work, ICT)
- People with emerging knowledge and skills (think: upcoming and on the horizon like robotics, artificial intelligence, nanotechnology).

It is essential to create and support a physical place for divergent people to converge to create inventions and innovation. One key factor in innovation is diverse elements coming together, and rural spaces (heritage and modern and high-tech skills) support a different kind of diversity from urban spaces (ethnic, multi-industry).[26]

Support could be provided to businesses and nonprofits to search for innovation and entrepreneurship in the region to support. Rather than waiting to be approached for funding, foundations and federal and state groups ought to seek out projects and businesses thriving and worthy of support despite the odds, similar to the approach taken by the United States Agency for International Development (USAID) in the early post–Soviet years in the former Soviet Union. USAID representatives went out into the community seeking projects to fund rather than waiting for applicants to come to them.

The federal government, state governments, and private foundations ought to fund asset-based qualitative interviews to uncover not-your-usual suspect problems needing addressing and not-your-usual suspect solutions being provided in the community.[27]

Governments, foundations, and universities must provide technical assistance for social entrepreneurs not frequenting the same social circles and loops as urban-focused private foundations or the federal government or universities. Some of the best ideas and projects in places hardest hit and most in need are by people not necessarily already in those circles.[28]

Support R&D Qualitative Asset-Based Assessments to Develop Potential for New Entrepreneurial Sectors.

The United States must devote university, R&D, and private research to create human-centered economies and human livelihoods in rural America for work that cannot be automated away.

Economic markets are created by humans and not extra-human forces.[29] With a loss of land knowledge through work not intimately connected to nature, often rural Americans do not know the value of what is underneath their feet. Exploring entrepreneurship and resource-based work not tied to historic or emerging large-scale single-sector industries in a rural site allows for the opportunity for work to develop that will not necessarily be automated away. With the brain drain and export of youth out of rural areas, local asset-based assessments and support for smaller-scale entrepreneurship offer the opportunity to potentially allow for more residents to stay who might otherwise be forced, for economic reasons, to leave.[30]

Programs and R&D must focus on return on small capital investments and modes of business mentorship and cooperation that do not require willing and hardworking participants to have college degrees.

Qualitative research and a regional publication outlet focused on the region can help the region discover its latent resources and its off-the-beaten path movers and shakers. We must support in-the-field discovery of our human resources and promote their connections across rural regions.[31]

Lose Rotten Rhetoric to Bridge Land and Identity.

The *us* versus *them* rhetoric of both the energy sector (against the EPA, the federal government, and environmentalists) and the environmental justice sector against the energy industry and its participants (surface mining, climate change, pipeline proponents) is toxic. The region had already been fitted for this rhetoric prior to the advent of Donald Trump's adversarial political rhetoric of *us* versus *them*.[32]

The allegiances of local people in coal country have shifted post-1980s from a union orientation to a company orientation, with the adversarial relationship with coal management replaced by an adversarial orientation to regulation.[33] The technological changes to the industry to reduce human worker need are absconded in this post–Reagan rhetoric. Models of land stewardship, linking to heritage and identity, bridging interests in rural life and hunting, fishing, and forest harvesting may offer ways to bridge the sympathies of local residents and heal rotten rhetoric. Agriculture and the region's livestock production history offer a path to the land connection that may resonate with remaining residents or attract new ones.

It is possible to be empathetic to the environmental and health results of people and the land due to coal production AND be empathetic to the miners and others who lose high wage work.

As the population dwindles in much of rural America with respect to the U.S.' growing urban population,[34] the United States must bridge urban and rural toward creating work and markets that cannot be automated away or off-shored and finding common points of identity and ethics we can abide together.

Engage in Thoughtful Energy Production in the Rural.

Coal, oil, and natural gas remain abundant U.S. resources.[35] It is unlikely they will be completely put out of use given their abundance in the United States and the world, especially in developing economies. For example, energy companies will continue to invest in technology to anticipate coal coming online again, but technology will replace human workers when it can be used.[36]

Natural gas is debated as to whether it is clean energy, as its carbon emissions as an industry rank just below that of coal.[37] Also, in the United States, coal is both a domestic and an export product, often moved through largely poorer communities by rail.[38] Again, as large commodity industries, energy industries will continue to focus on technological solutions to access, production, transportation, and so on in lieu of hiring workers.[39] While large corporations control millions of acres of land for energy production, they are not going to hire millions of workers. Given the abundance of coal and natural gas resources, companies are unlikely to let go of land they already hold or hold mineral rights to. Rural planning and research must assess and track these companies, landholding companies, and industries. Local planners must keep abreast of the local energy mix, its players, and what role they anticipate continuing to play in the local rural industrial space.

Refocus Public Universities and Their R&D on the Public Good.

The fossil fuel industry places tremendous pressure on research universities (benefactors of their largesse) and their economic development departments to develop technology that will allow the sustainable and clean use of fossil fuels, and the commodification of timber, water, and other resources both directly for use in large-scale industry and by the government (institutions and military) and indirectly through externalized industrial costs (processes developed to allow industry to have their environmental impact or training or retraining costs paid by taxpayers rather than directly by industry).[40] Both corporate and federal contracts and grants (such as through the Department of Energy and the Department of Defense, the latter being the U.S.' greatest consumer of energy) offer tremendous support and incentive for this devotion of research foci and economic development department resources.[41]

Cheap energy provides the backbone of our current economic model and enables nearly all our other current economic sectors.[42] That energy remains cheap and plentiful is not only an industrial and economic imperative but also an internal and external security one.[43] For example, the interest in central Appalachia's abundant energy resources will always include national and international interests. The following list, drawn from the internet c. 2014 (West Virginia Development Office, and no longer available online, and no updates found as of this publication), illustrates the international character of West Virginia's international investment, and, many of the names of the businesses spell out those interests outright: "coal," "alloy," "pulp," "steel." That this range of international investment in a resource rich rural area comes as surprising underscores how skewed popular and even academic understandings are of economic ties and interests in what seems to otherwise be "pastoral" or "natural" settings. We "black box" rural places, our understanding of their complexities remaining immature and inaccurate:[44]

Argentina: Unitan
Australia: ALS Environmental, Inc.; Incitec Pivot Ltd./Dyno Nobel Minova
Austria: ANDRITZ Separation Inc.; Virginia Crews Coal Co.
Belgium: Proviron Industries; Ravago
Brazil: Braskem SA; Gerdau Ameristeel; Madem Reels USA
Britain: AFAC; Alamo Energy Corp.; BG Group; Caledonian Alloys; Elementis; Ferguson/Wolseley; International Power PLC; ProLogic; Randoz Laboratories; Rio Tinto; Smallbone; Thistle Processing
Canada: Bombardier Services; Bookfield Power; Chrome Deposit; Cobalt Coal; Hercules Inc.; H Q Aero Management; Irving Crane; Kent Cartridge of America; Leveltek; MAAX; Mott Manufacturing; North American Ventures; Quebecor Printing; SFK Pulp Recycling; Stella-Jones; Trimac; Western Inventory
China: GT Global/The Daniels Co.; Phillips Machine Service, Inc.; Terramite Construction Equipment
Denmark: FL Smidth
Finland: Mesto Power Service
France: Axens; Constellium; GDF Suez Energy; Saint-Gobain/CertainTeed; Saint-Gobain/Corhart; Saint-Gobain/Norandex; Sonepar/Hagemeyer NA; Veolia Environmental Services
Germany: Baum America Corp.; Bayer Crop Science; Bayer Material Science; Becker Mining Systems; Brenntag Mid-South; Century Lubricants; Fasloc; Fontaine Engineering; FLYTEG North America LLC; Kaeser Compressor; Kloeckner-Pentaplast; Marco North America; Mato; Prebena N. American Fastener Co.; Schauenberg Flexadux Corporation; Siemens Medical; Siemens Water Technologies; Seebach; Seetech; Stockmeier Urethanes Inc.; Tiefenbach; ThyssenKrupp Elevator; ThyssenKrup Safway; Vossich Track Material, Inc.

India: Essar Group; Novelis
Ireland: Airdagh Group; CRH-L/Oldcastle Precast
Israel: ICL/Clearon; ICL/IP
Italy: Allevard Sofegi USA, Inc.; American Agip Co. Ltd.; DPR, LLC; Entsorgia; IMI Fabi; Italcement/Arrow Industries; Italcement/Essroc; M & G Polymers; Pietro Fiorentini USA; Tecnocap LLC.
Japan: Diamond Electric Mfg.; Feroleto Steel Co.; Green Metals; Hino Motors; Kuraray; Kureha PGA; K.S. of West Virginia; Melji Corp.; Mitubishi Rayon/Lucite International; NGK Spark Plug Mfg.; Nippon Thermonstat; Nippon Tungsten USA; Okaya; Okuno International; Sanko Denki; Taiyo Nippon Sanso; Teikoku USA/Chempump; Toyota Motor Mfg.; Toyota Tsusho; Wheeling-Nisshin Inc.
Kingdom of Bahrain: Armacell
Luxembourg: Arcelor/Mittal/Concept Group; Flint Group
Mexico: Grupo Bimbo
Netherlands: Impress USA; Royal Ahold; Shell WindEnergy; Shell/East Resources, Inc.
New Zealand: Rank Group/PWP
Norway: Statoil Energy
Republic of Korea: Gastar Exploration
Russia: Mechel OAO; Mountain State Carbon
Saudi Arabia: SABIC
Spain: Gestamp; Maxam North America
Sweden: ABB Process Analytics; ABB Service
Switzerland: Aggregate Industries; EuropTec USA; Holcim; Novartis/Alcon Research, Ltd.; SGS; Sulzer Pumps/Sturm
Ukraine: Felman Production; Metinvest/Carter Roag Coal; Metinvest/United Coal

Support Work That Cannot Be Automated Away.

To insist upon the refocusing of regional and rural research and economic development resources toward supporting economic sectoral development in which workers will not be displaced by technology is not to demand some Luddite throwback to another era.[45] It is but a shift in the goals of efficiency and technical direction, and service. Nearly every research university in central Appalachia is a public university, and, with significant taxpayer support and tasked with developing work and workers for that region's economic future, thus also ethically bound to serve the region's people. This means work for generations, rather than work that will inevitably, sooner rather than later, be displaced.

This same region's university-based research ought to support economic sectors and develop work as well as financial, political, and community processes that will not displace workers with technology, and, that will address land, water, and community stewardship imperatives. This kind and quality of support is also possible to incentivize through federal grants and corporate contracts.[46]

Give It Back.

If legacy is key, give control of nonindustrial lands or land not in use back to the original indigenous inhabitants. Arguments, academic or otherwise that Native Americans did not inhabit the region of the current Pocahontas Coalfield have been disproved, and, even if they had not, then, other adjacent and proxy remaining groups can be found.

Honor Homegrown and Connect Homegrowns.

Models that are homegrown initiatives with much local buy in and are community-derived and disseminated should be connected across spaces for resource and lesson sharing.[47]

Offer Dignity and Hope.

One local social service provider respondent highlighted needing projects and programs that restore dignity as a top priority; among so much loss, pain, and destitution, finding a way to contribute and to be valued is key.[48]

Scale Appropriately.

Projects ought to focus on appropriately scaled ways to diversify their local economies with a focus on local assets both in material goods and in social capital.

Provide Cover.

Part of the culture is to doubt the opportunity for less familiar modes of doing things to succeed. Social entrepreneurship models also can provide a good "cover" for local folks to try new businesses, approaches, and so forth.

Support High Potential for Innovation Diffusion.

Focus on models with success that are researchable and with results that are verifiable. Thus, lasting models can be dissected and replicated for regional diffusion.[49]

Tell New Stories.

Focus on narratives that shake up the single-sector narrative and provide meaning beyond the single sector. Focus on practical work that sidesteps automation.

Plan and Account for Low Population.

Though many other issues face rural spaces, population decline strikes the hardest blow. Simply, without population, you cannot fill jobs, start or support local businesses, fund schools, and compete for private and federal grants if you are following a model for competition with urban space, a model favoring consolidation, a model of endless growth.[50] Likewise, the low and declining population in much of rural America handicaps business and nonprofit competitiveness and nonprofit measures of quantitative impact.

Low and declining population will remain a key struggle. The United States needs a national discussion of "right size" for the rural, or, how many people an economic sector in a place truly can support. For example, from a rural swath of the southeastern US from Montgomery County, VA to McCreary County, KY across fifty counties roughly along the 36th parallel, the sum of that population is less than that of metro Pittsburgh.

As the rural population continues to decline, we must find ways to bridge sectors across state lines and to bridge interests. There are simply not enough local residents in sections of some individual states for economic sector participants to compete across state lines. Everyone must be brought to the table and engaged. Many inter- and cross-regional alliances across economic sectors must be jumpstarted to produce economic viability for the large rural sections of our region.

Economic policy ought to shift toward smart shrinkage, low growth, and even potential degrowth. In short, most of this kind of rural America is losing its population, and becoming increasingly poorer, and regional and national rural policy ought to address this head on.

We must have a frank discussion in the United States more generally about population decline in rural areas. We must frankly and openly plan for the shrinkage of the population in most rural areas.

Europe spent much time in the early 2000s rethinking its approach to its population declining areas. The European Union developed the Shrink Smart program and only one U.S. city has implemented a similar program (the newly "Appalachian" town of Youngstown, OH). Likewise, the World Bank made a series of recommendations for the relocation or phasing out of small populated Russian monotowns.[51]

In the United States, we remain well behind on this important set of policy prerogatives. In an elegant discussion and table, Martinez-Fernandez and Wu (2009) (see www.bigrural.space for the full graphic), lay out the conceptual frameworks for rethinking a city's purpose in an age of shrinkage, and now, Post–COVID-19 shifts (until the next pandemic or global crisis hits). Highlights of their succinct summary of "Urban & Regional Policies & Strategies" cross-referenced with major drivers for cities shrinking are also worth summarizing here.

Policies addressing "Industrialization Center-Periphery" issues include growth at both metropoles and centers, and, "Decentralization policies aimed at stimulating growth in the periphery" (read, rural, beyond the commutershed). "Deindustrialization and Post-industrialization" policies may include urban renewal programs, tax free zones, industry clusters, and "smart cities, 'creative cities,' 'design cities.'" Sectoral studies regarding the rural address climate change through "water management; disaster recovery" and confront directly coming "climate refugees."[52]

In this concept of shrinking spaces and places, we must come to face the fact that some subsections of our region may not be jump-started economically. We must then discuss what is required to live in these places: what should these kinds of communities' "survival and thrival" tactics become?

While we grow these emerging economic sectors, how can we re-skill for homestead and rural livelihood in locations where even new emerging sectors may not take root, or root deeply enough? How do we offer skills to make rural living possible for people wanting to stay in their location, but in a generation not raised to live from the land?

We must also begin to plan now for the day the relief stops, for both sectors and for individuals.

Develop Leaders/Leadership.

Low trust often exists among the rural population for their elected leaders. Alternatively, few thriving centers exist to train residents in civic leadership or participation or to encourage or train locals to run for state or local office. Make rural leadership training imperative.[53]

Create Leadership Institutes and Programs for Children and Adults at All Levels of Society in the Rural United States.

Many texts cite long standing issues of corruption and cronyism in rural single sector regions. Current qualitative interviews suggest this is the number one issue local citizens wish to address, as, without leadership, emerging economic sectors will falter, educational institutions will fail, and the local population continues to lose hope.

Coordinate Public Health Response.

Large swaths of rural America are targeted by the Centers for Disease Control for their explosions of HIV, Hep B, and Hep C due to the opioid and other crises of addiction. These outbreaks also follow historically marginalized or rural single sector areas.

Currently, little deep and large-scale federal coordination of the response to rural HIV, Hep B, or Hep C exists, and, likewise, little extra-statutory coordination of response to these diseases. As the issue of poor health looms so large in much of the rural United States, with a long list of chronic diseases ranking high, the coordination of public health response must be key.[54]

Coordinate Education.

Little to no extra-statutory (cross-border) coordination exists in rural America pertaining to economic sectors in which work has less of a chance of being automated away. The education that does exist is not tied to providing appropriate entry point education for novices or the unemployed as well as professional development for experienced professionals in emerging sectors. Furthermore, education for novices often is not tied to market imperatives or predicting upcoming work trends.

Fund Regional Educational Coordination for Emerging Regional Economic Sectors.

Create regional educational coalitions and conferences to address the range of emerging training needs for the emerging economic sectors. Centralize a system for credit participation, where students may earn credit or degrees through credits earned at various locations. Support a range of regional initiatives addressing professional development.

Investors and Advocacy: Finance and Government Finance.

Rural America needs investment scaled to local needs, and, it needs to link investors to each other for education on impact investing for emerging economic sectors to succeed. If federal and state entities do not step up to assist with this emerging economic sector organization, then the private, nonprofit, philanthropic, and education sectors must step up to do this work themselves. This governmental assistance so far largely has been lacking and not comparable to the assistance given to energy or large-scale industry; moving forward we can anticipate that this will remain true.

We must move forward from a model that more than likely, in many locations, who is left and what is left *is* what is left to be built on.

We can build on how emerging economies and people already flow in rural regions. We can follow how people already currently network (such as media markets, transportation flow, even how schools play sports against one another) to form connections in rural subsections that make the task of working together manageable and sensible.

Private foundations and federal shifts need to scale grants that make sense for low rural populations and/or to offer technical assistance to help counties work across borders to serve higher population impact.

The giving to rural areas does not match, per capita, the giving to urban regions. Federal and state grants are tooled for quantitative impact (how many people served). New kinds of federal and state and private foundation grants must be created to address the scale and needs of rural giving. Likewise, appropriate investor giving ought to be scaled and created for rural regions.

Philanthropic, governmental, and private investment must support giving and lending in rural economic sectors that will not lead to the automation of rural jobs and the continued displacement of rural workers by technology. The U.S. Bureau of Labor recently anticipated that many jobs currently paying less than $20 will be automated in the future.

Dump NIMBY (Not in My Backyard).

Very plainly, whether offshore, abroad, or in the rural United States, the bulk of consumer goods as well as products resulting from large technical systems (energy, for example), have been situated in the U.S. outside metropoles. The rural has become an “out-of-sight” industrial resource area for society’s centers. Along with increasing food and energy security, co-locating food and energy production in urban, suburban, and exurban environments can diversify those sites' economies, reduce the Ag and energy burden from the offshore, provide opportunities for entrepreneurship, and reintroduce metropole dwellers to the sources and human, environmental, logistical, and energy complexities of providing of their food and energy. If food and energy production are safe and not unsightly, there is no reason for them not to be co-located with metropole dwellers. We anticipate technology and science R&D to answer the current paradigm of offshoring/NIMBY-ing food and energy production. Why not then support significant science and technological research for bringing energy and food production down to a smaller entrepreneurial scale with wider and broader distribution and disbursement?[55]

Focus on Soil and Water Health.

Without healthy soil or healthy water, we cannot build on an agricultural or pastoral past for a lively and autonomous future. Coordinating across state borders and developing agricultural and land use protocol and education across rural regions for soil and water health remains key to producing entrepreneurs who can shepherd both our rural land and economy into the future.

We must seriously consider unpredictable weather as a factor in the kinds of long-term enterprises we seek to steward, and we must plan and respond to the demands of highly erratic weather.

Rural America and Its Organizations Must Address the Elephants in the Room.

Identity

Our relationships to each other and to place evolve. For example, prior to the rise of coalfield identity in central Appalachia in the 1880s, people in subsections of this region identified in various ways: Southerners, pioneers, sharecroppers, with their families or kin, with their professions, with their states, and so forth. Prior to this, Europeans in the region were trappers, surveyors, or soldiers, and the people who worked on and lived on, and loved, this land prior to European involvement were mostly gone from the region, removed through death, conquest, and war, or by forced migration. Thus, an "Ur" (proto, primal, or original) claim to Appalachia by many current residents highlights the complexities and even complications of place-based identity.

Treading in and out of much central Appalachia place-based identity also is a strong masculine work identity: a coal mining family is defined as such because the men in that family work/worked in the mines. Every family member becomes part of the mining technology that is the coal miner himself.

Identities shift and change, and we can redefine relationships and regions to suit our current emerging modes and participants in work. We can come together to articulate a common vision, common mission, and common values with respect to our self-descriptions and our current and future stewardship of land for which we find ourselves responsible either by choice or by industrial overdevelopment and abandonment. We can become a sustainable agriculture region, or as a farmer in Greenbrier County, WV wished West Virginia to be, "a new Vermont." We can become a cultural, liquor, environmental, recreational, and farm tourist region. We can become a remote work region.

We can redefine our region and make new subsections through new connections or rekindling broader or older ones. We can embrace shrinkage, degrowth, and "right size" as positives.

Regions ought to come together to form a common purpose and common vision for this common ground through means such as a regional envisioning summit, branding and marketing campaigns, theorizing through public intellectuals and grasstops, and so on.

Just as these prior "Appalachian" identities were not forever, other rural American identities are not forever. How else could or ought political, personal, economic, and cultural connections from rural regions or subregions be constructed?

Race and Class.

Race and class remain large dividing factors in many rural communities. Very rural sections of the United States with African Americans, Latinos, and Native Americans receive less state funding and less national attention, and African Americans and groups of other races often remain invisible at academic, federal, state, nonprofit, and other conferences, in media in the region and beyond, especially lacking recognition for and in positions of leadership.[56]

The underclass (in terms of earnings) of rural America continues to be a target of mixed local feelings, especially by a region's remaining middle and upper classes.[57]

Open and frank discussions must occur regarding the difference between a mode of rural living focused on homesteading and self-sufficiency, production for export, the continued shrinking of local economic purchasing markets, and the models of continued economic relief.

Lesbian, Gay, Bisexual, Transgender, Queer, Intersex, Asexual (LGBTQIA).

Just as race and class divides exist, the recent spate of marriage inequality attempts and restroom gender laws create a hostile atmosphere for the *LGBTQIA* community. For rural regions to prosper, or even not dwindle away, they must engage all their citizens.

Programs aimed specifically at supporting the rural *LGBTQIA* community ought to be funded. In the profile of foundation donations, gifts that support diversity ought to be highlighted and promoted. The economic power of folks identifying in this group also ought to be highlighted and promoted as a potential regional driver. Last but not certainly least, all U.S. citizens deserve to be treated equitably, be free to move as they please and, under the U.S. Constitution, live where they want, including living in and being protected as equally as other citizens in the rural.

CONCLUSION

Big Rural can be dismantled. Big Rural is not too big to fail. A national rural strategy could be a foundational instrument toward breaking apart Big Rural. Of course, a strategy or policy stands not as an end-all or be-all with respect to policy or action in the polity. As a reviewing scholar reminded me some years ago, the People's Republic of China has liberal freedom of expression laws on the books. Any policy or strategy proves only as good as the institutions tasked or taking responsibility for enacting it. The lens for this National Rural Strategy included a focus on research and development for support for emerging or increasing sustainable economic sectors in the rural United States, with nods to, education, entrepreneurship, rural innovation, asset-based assessments, energy stewardship, health, and identity.

Scientists, engineers, financiers, academics, and government officials all play significant roles toward enabling a robust rural strategy that supports work that cannot be automated away and disburses more democratic technology in major areas of production usually out of sight of metropole consumers. Their policies and strategies with respect to scientific and scholarly research agendas can inhibit or promote liberty and equity. Scientific and technical practice as knowledge seeking affects the polity, in the lived environment and experience of citizen-residents. In this current chapter, I propose means by which research can actively support more liberty and equity. I define concrete practices that ought to lead to a more equitable distribution of the benefits of science and knowledge-making by outlining how universities can engage with and on behalf of the citizens they ought to serve. I link specific practices to the major issues I have argued in the subsequent chapters: automation, democracy deficit in single sector economic work-civic totalitarianism, the flattening of metis by scientific and technical practice and machines, and brownfield rehabilitation.

NOTES

1. The U.S. Census sets the line for rural versus urban at 50,000 inhabitants. 2010 Census Urban Area FAQs," n.d. https://www.census.gov/programs-surveys/geography/about/faq/2010-urban-area-faq.html; National Security Strategy 2010. Administration: Barack Obama. Executive Branch, n.d.

2. Bureau of Labor Statistics, Frey and Osborne, 2013, CEA Calculations; "Artificial Intelligence, Automation, and the Economy." Washington, DC: Executive Office of the President, December 2016. https://obamawhitehouse.archives.gov/sites/whitehouse.gov/files/documents/Artificial-Intelligence-Automation-Economy.PDF

3. A major point made by popular writers such as Daniel Pink, however, these same concerns can be extrapolated from documents emerging from the Executive Branch in 2016. See Kristin Lee, "Artificial Intelligence, Automation, and the Economy (blog)," December 20, 2016, Washington, DC: Executive Office of the President, December 2016. https://obamawhitehouse.archives.gov/sites/whitehouse.gov/files/documents/Artificial-Intelligence-Automation-Economy.PDF

4. Karina Pallagst, et al. ed., *The Future of Shrinking Cities: Problems, Patterns and Strategies of Urban Transformation in a Global Context* (Berkeley: Institute for Urban and Regional Development, Center for Global Metropolitan Studies and the Shrinking Cities International Research Network, 2009).

5. Cristina Martinez-Fernandez and Chung-Tong Wu, "Shrinking Cities: A Global Overview and Concerns about Australian Mining Cities," in Pallagst et al., *The Future of Shrinking Cities*, 29–36.

6. David Leadbetter, "Single-Industry Resource Communities: 'Shrinking,' and the New Crisis of Hinterland Economic Development," in Pallagst et al., *The Future of Shrinking Cities*, 89–100.

7. For example, in November 2017, China Energy Investment Corp. proposed investment in West Virginia-based natural gas. *Reuters* Staff. "China Energy Investment Signs MOU for $83.7 Billion in West Virginia Projects," Reuters, November 9, 2017. http://www.reuters.com/article/us-trump-asia-energy-west-virginia/china-energy-investment-signs-mou-for-83-7-billion-in-west-virginia-projects-idUSKBN1D90S9?il=0.

8. Merrit Roe Smith & Leo Marx, Does Technology Drive History? *The Dilemma of Technological Determinism* (Cambridge, MA: The MIT Press, 1994).

9. Respondent, 2016.

10. Artificial Intelligence, Automation, and the Economy, 2016. https://obamawhitehouse.archives.gov/blog/2016/12/20/artificial-intelligence-automation-and-economy

11. For information on The Highlander Center's Appalachian Transition Fellowship, which ended in 2019, see: https://highlandercenter.org/programs/appfellows/.

12. Erling Roed Larsen, "Escaping the Resource Curse and the Dutch Disease?: When and Why Norway Caught Up with and Forged ahead of Its Neighbors," Statistics Norway, Research Department, 2004. http://hdl.handle.net/11250/180569.

13. Timothy Mitchell, *Carbon Democracy: Political Power in the Age of Oil* (New York: Verso, 2011).

14. Rebecca Scott, *Removing Mountains: Extracting Nature and Identity in the Appalachian Coalfields*, Quadrant (Minneapolis: University of Minnesota Press, 2010); Shannon Bell, "'There Ain't No Bond in Town Like There Used to Be': The Destruction of Social Capital in the West Virginia Coal Fields," *Sociological Forum* 24, no. 3 (September 2009): 631–57.

15. APPENDICES: Further Reading.

16. Respondent, 2016.

17. Author's field observation, 2016.

18. Pallagst, Karina, 2009.

19. The U.S. currently provides much R&D for private sector development, particularly in the energy sector through the Department of Energy, Department of Defense, and National Science Foundation.

20. Christopher D. Barton, Professor of Forest Hydrology and Watershed Management at the University of Kentucky, in discussion with the author, July 2016.

21. Steven High, "Capital and Community Reconsidered: The Politics and Meaning of Deindustrialization," *Labour/Le Travail* 55 (2005): 187–96.

22. Author's field observation.

23. In 2014, for example, an academic database search brought me to one academic paper from the 1980s on heroin in Appalachia.

24. Jerry Bruce Thomas, *An Appalachian Reawakening: West Virginia and the Perils of the New Machine Age*, 1945–1972. Vol. 12 (Morgantown: West Virginia University Press, 2010); Rebecca Scott, Removing Mountains.

25. Freudenburg, William R., "Addictive Economies: Extractive Industries and Vulnerable Localities in a Changing World Economy," *Rural Sociology* 57 no. 3 (Fall 1992): 305–32, https://doi.org/10.1111/j.1549-0831.1992.tb00467

26. Mark Muro and Bruce Katz, *The New 'Cluster' Moment: How Regional Innovation Clusters Can Foster the Next Economy* (Metropolitan Policy Program. The Brookings Institution, September 2010).

27. "Master of Social Work," n.d., Graduate Programs, Concord University, https://www.concord.edu/academics/online-graduate-programs/master-of-social-work.

28. Field Observation, 2016.

29. Michel Callon, ed., *The Law of the Markets* (Hoboken, NJ: Wiley-Blackwell, 1998).

30. Patrick Carr and Maria Kefalas, *Hollowing Out the Middle: The Rural Brain Drain and What It Means for America* (Boston: Beacon Press, 2010).

31. David Ahlstrom, "Innovation and Growth: How Business Contributes to Society," *Academy of Management Perspectives*, 24(3), 2010, 11–24; Joe Tidd, John Bessant, and Keith Pavitt, *Managing Innovation: Integrating Technology, Market and Organizational Change*, 3rd Edition (Chichester, UK: John Wiley & Sons Ltd, 2005); Johann Fueller, Hans Muehlbacher, Kurt Matzler, and Gregor Jawecki, "Consumer Empowerment through Internet Based Co-creation," *Journal of Management Information Systems* 26, no. 3 (Winter 2009–2010): 71–102; Eric Von Hippel, "Horizontal Innovation Networks-By and For Users," *Industrial & Corporate Change* 16, no. 2 (April 2007): 293–315.

32. Michael Grunwald, "Trump's Love Affair with Coal," *Politico*, (October 15, 2017). https://www.politico.com/magazine/story/2017/10/15/trumps-love-affair-with-coal-215710.

33. Rebecca Scott, *Removing Mountains.*

34. Surprisingly, the total U.S. rural population has held steady over the last sixty years. However, it now represents a much smaller piece of the total US population: https://data.worldbank.org/indicator/SP.RUR.TOTL. Note 54,208,527 in 1960 and 56,378,923 in 2022.

35. Susan Tewalt, Jason Willett, and Robert Finkelman, "The World Coal Quality Inventory: A Status Report," *International Journal of Coal Geology* 63, no. 1/2 (July 2005): 190–94.

36. World Nuclear Association, "'Clean Coal' Technologies, Carbon Capture & Sequestration," (updated November 2021). http://www.world-nuclear.org/information-library/energy-and-the-environment/clean-coal-technologies.aspx.

37. Paulina Jaramillo, Michael Griffin, and Scott Matthews. "Comparative Life-Cycle Air Emissions of Coal, Domestic Natural Gas, LNG, and SNG for Electricity Generation." *Environmental Science & Technology* 41, no. 17 (2007): 6290–96.

38. Union of Concerned Scientists. *The Natural Gas Gamble: A Risky Bet on America's Clean Energy Future* (March 4, 2015). http://www.ucsusa.org/clean-energy/coal-and-other-fossil-fuels/natural-gas-gamble-risky-bet-on-clean-energy-future#.WeMjKxmGPrc.

39. Travis Lowe, "I've Seen the Future. It Looks Like Appalachia: Life after the Job Apocalypse," *Medium* (March 31, 2017), https://medium.com/startup-grind/ive-been-to-the-future-it-looks-like-appalachia-5553bf89f275; Respondent 2016; Author's field observation; Ed Felton and Terah Lyons, *The Administration's Report on the Future of Artificial Intelligence* (blog), October 12, 2016. https://obamawhitehouse.archives.gov/blog/2016/10/12/administrations-report-future-artificial-intelligence.

40. "University Training and Research (UTR)," Office of Fossil Energy and Carbon Management (FECM), updated to include 2018 information, https://www.energy.gov/fecm/university-training-and-research-utr-program

41. Moshe Schwartz, Katherine Blakeley, and Ronald O'Rourke. Department of Defense Energy *Initiatives: Background and Issues for Congress*. Report prepared for the Congressional Research Service, December 10, 2012.

42. Institute for Energy Research, "North American Energy Inventory," December 2011. https://www.instituteforenergyresearch.org/wp-content/uploads/2013/01/Energy-Inventory.pdf.

43. Congressional Research Service, "Department of Defense Energy Initiatives: Background and Issues for Congress," by Schwartz, Moshe, Katherine Blakeley, and Ronald O'Rourke. June 26, 2012.

44. http://www.wvcommerce.org/App_Media/assets/doc/businessandworkforce/Intl/Investment_Flag_Map_2014.pdf

45. Felton and Lyons. The Administration's Report on the Future of Artificial Intelligence; Lowe, "I've Seen the Future."

46. Jacob Demmitt, "Virginia Tech Planning 300-Acre 'Automation Park' to Test Drones, Self-Driving Cars," *The Roanoke Times*. August 29, 2016. http://www.roanoke.com/business/news/virginia-tech-planning—acre-automation-park-to-test-drones/article_56e245ff-af92-54a4-91ac-da839cc327f2.html.

47. Respondent.

48. Respondent.

49. Eric von Hippel, *Democratizing Innovation* (Cambridge, MA: MIT Press, 2005).

50. http://www.shrinksmart.eu/

51. World Bank, 2011.

52. Martinez-Fernandez, Cristina and Chung-Tong Wu, "Shrinking Cities: A Global Overview and Concerns about Australian Mining Cities," in *The Future of Shrinking Cities: Problems, Patterns and Strategies of Urban Transformation in a Global Context*, ed. Karina Pallagst, et al. (Berkeley: Center for Global Metropolitan Studies, Institute of Urban and Regional Development and the Shrinking Cities International Research Network, 2009), 29–36.

53. Shannon Bell, "There Ain't No Bond in Town Like There Used to Be"; Ledet, "Correlates of Corruption."; Respondents.

54. Respondent.

55. The geographic profile of a rural area often dictates its utility to the metropole. For example, in 2013 I brainstormed and then reviewed purpose and work in the rural seeking to find patterns. I identified remoteness from the metropole as a key feature in the following sectors/facets in the rural:

- mining
- industrial agriculture
- oil and gas
- recreation
- hydro production
- aquaculture
- fishing
- industrial transportation
- shipping
- energy production
- water production
- wastewater management
- waste management & storage (individual and industrial)
- college towns
- factory towns
- bedroom communities
- tourist towns
- tourist sites
- national forests
- chemical factories
- paper mills
- retirement communities
- military training
- hobby farming
- timber production
- subsistence farming/homesteading

- intentional communities
- artisanal farming
- religious communities
- transportation hub/logistics
- incarceration/prisons
- mental institutions
- relocation
- deindustrialization
- demise
- no growth
- degrowth
- shrink smart
- economic diversification
- power generation
- subcultural haven
- creative economy
- drug addiction and sales
- exporting youth
- Federal and/or state dependency
- Relocalization of skills
- Skill sharing
- Community subsistence practice
- Barefoot College
- Invention center a la Open Source Technology/MakerSpace

Training people for skills that ensure human capital export.

56. Respondent. Field Observation.
57. Respondent. Field Observation.

Interlude 3

Amelia Bandy

In *Big Rural* we have detoured through interviews with grasstops to the history of technology to the philosophy of technology to failure of policy to the assertion of strategy, all while keeping foregrounded the democratic citizen, the state of the citizen in the single-sector rural industrial space.

The order of interviews in this volume finds some logic: Jason takes us deeper into the journey and circumstances of a place like McDowell County, WV, Atlas sets some of the theoretical stages for this kind of rural industrial space, and Amelia brings us back again to the structural, the path of collaborative work and its implications. Their "takes" weave together a fiber of sought after dignity for rural people.

Amelia has worked in public health in the region and works to reintroduce mountain farming practice, sector support, and people support in McDowell County, WV and Tazewell County, VA. Amelia and I were talking across her dining room table near Richlands, VA.

Amelia: Most of my family is still here and they live close by. That is probably a big reason as to why I am still here. Like any Appalachian, I have parts of my family that are complicated, and layered. This all informs the work that we are doing and the decisions we have to make. We have so many things going on in our families that can make it difficult to navigate.

My grandmother is still alive and she mostly raised me. My parents are still alive, and working on their relationships. They are divorced. We have different members of our family where people are raised together with their parents—a child being raised as a sibling. The truth is not told about what is going on till you are much older. My family has had issues with pharmaceuticals. This adds a difficult layer to having to care for your family. This is what is going on with many of the families in this area.

People not from this area have a picture of what using drugs looks like or someone that would overdose. In a lot of cases, it does not look like that at all. It makes it difficult to contend with because it looks like it's normal, it's all okay. I have had to ask myself—someone from another organization asked me why I am still here. It was weird to have a group that wanted to work with us ask me that. I have had to ask myself that, too. There have been times when I wanted to leave, but I never had a major plan. I tend to want to fix things. When I try to think about it, even in my earlier positions for employment, I was doing work in the community—offering programs, education, and awareness.

Crystal: So, pick me a story in there. Let's get past the more "newspaper account." Let's get to it. You are raising a biracial daughter here. The thing I contend is that race is the elephant in the room down here.

Amelia: My daughter is seventeen. Raising a child is difficult. Period. And being a single parent is difficult. Raising a biracial daughter in this area is difficult. And though she looks like me, people do not care that a White mother is raising her. It does not gain her any favor in this world that her mother is White.

In this area, as she has gotten older, it has become glaringly apparent that racism is still very much alive. We have had so many challenges—it's really heartbreaking. I am trying to raise her to be all the things you would want your child to be, and there is this other layer, like, she gets out of the car and goes into a store by herself, I have to look at her and say, keep your antennas up. I would not have to say that to her if she were White. Obviously, you want your child to be safe. Period. There are experiences that she is going through that I don't believe would go on if I were raising a White child or raising her in a more diverse area.

I will be right beside her somewhere and see an older person look at her, and she has had people look at her, like, what are doing here? When you have this, it makes you question every single interaction she has: is whatever happening because she is Black? There are so many things that happen to her. It is nearly a daily thing. From people trying to touch her hair like she is an object on display to asking her questions about her hair or whatever choices she makes.

She texts me nearly daily with some round of bullshit about what it is like for her, and I don't know if she ever gets the opportunity to really feel free. She has to think about things in a different way than even her friends do. It is becoming important to her to understand who she is and where she comes from. I can't teach her these things. We read books together—we are reading *Before the Mayflower* now. There is a history of her and her people before slavery that she will never hear about in school.

Her father is in prison and has been for most of her life. He wants to be in her life, but maybe may not know how. Some of his family is involved with her. There is a whole side to herself she may not even know. She has had issues with school since elementary, where she has asked, could you please teach Black history other than in February and something also beyond Martin Luther King, Jr.? I want these people to understand something of my history. Their answer to this was to give her a book and tell her she could recite a poem out of this on the announcements. She was like, what? There are parts of this I will never understand or identify with. If I express anger or try to get to the bottom of things,

I have had White people say to me that they think Black people make that stuff up. There is something that they started doing at her school for spirit week: redneck day versus country club day. There are people that are proud to be country; some people are getting made fun of because they don't have much, why would you celebrate or differentiate about that at school?

Crystal: Making fun of people you go to school with. And this would be one thing that characterizes Tazewell County, VA from McDowell County, WV is that there is a country club set in Tazewell. There was also a plantation derived set of folks here. There has been a large farm set here. There was money prior to coal. And there is also now coal money in Tazewell. There is already that tension between country and country club.

Amelia: She goes to school that day and there are people wearing full on Rebel flag coats, hats. Someone was wearing a Rebel flag like a cape. She calls me and says this is all making her uncomfortable, that she did not want to be at school.

So, I go to the school. The assistant principal comes, and I am trying to explain to her, that my daughter and her best friend, who is also Black, and they do not feel safe here. They were trying to understand what this is all supposed to be celebrating.

She says to me that those students don't think what you think they mean. And I respond, I don't really care what you think or they think, I am telling you that my daughter and her friend, how they feel. That should be important to you. I know there are layers to this thing. Maybe they don't relate it to race. This is so normal here.

Crystal: I know that wearing Confederate insignia to school is currently legal in Virginia, but there is legal and there is what is right.

Amelia: I don't always relate to folks in the White community here. They don't relate to me or think I am making something out of nothing. Folks in the Black community are like, yeah, of course, we have experienced this.

Crystal: When I grew up here, there were a lot of different kinds of people living here.

Amelia: Even when I think about some of the work we did in the beginning [community work], coming together for the working groups, part of my role was to soften the blow that we were working with Black farmers in Appalachia. We weren't taken seriously.

Amelia: It is difficult to tell personal stories because you have to reveal things about yourself. But, with the producers we are training [in mountain farming and entrepreneurship], they have difficult things in their backgrounds, and it is easy for me to see where they are coming from. I share some of the experiences they've had. I have not personally had a drug problem or been incarcerated or in the military or in a war, some of the things our folks have gone through or are going through currently, but I am not too far separated from them.

Some of my decisions could have landed me there. We've all made bad decisions, and we are really only one bad decision away from being in that person's place. In trying to build what we are trying to build, we have to have a different layer of understanding. It's not cut and dry. We have had people go through our program that are in recovery, or maybe in an active state of addiction. You have to have a level of understanding of where they are coming from.

Crystal: I don't find the truth mortifying. I don't find the complications of my family and what they went through mortifying because I have traveled and know that what happens here, happens everywhere, by everyone and to everyone, and we are not exceptional in these complications here in Appalachia.

Amelia: The region did not get to the low point it is in a year—not just in terms of coal. This area has been attacked, and either we fight back or, there is no or. We have to do this. I have been affected by family members that are addicted. I understand what that is like. I also understand that there is a missing component in our region. The family unit has been attacked. The community has been attacked.

Crystal: People will pick on that word "attack." Tell me more about that.

Amelia: We had pharmaceutical companies target this area because folks had jobs working in the mines and needed pain relief. Folks were told nothing about what was going to happen to them. They dumped this stuff into these places.

Crystal: We were the guinea pigs for this. I have told you about how my elderly mother was put on low-dose opioids, and the doctor told me that they were harmless. I saw that they made her goofy. She was terrified of being in pain. She went to visit someone, and a houseguest stole the pills. She could not get a new prescription for four or five days, so she began withdrawals—with all the ones you have from getting off opioids. I called her doctor, as I did not want her to be in pain and I wanted to let him know that a bottle of pills with his name on it was now in someone else's hands. I asked him, so, what could she take instead to help with her pain. He responded: Extra-strength Tylenol. I was dumbfounded. I said: You mean to tell me that all these years my mom could have just been taking Tylenol? He was silent on the other end. The opioids did not kill my mother, but they greatly diminished her quality of life the last ten or so years of her life. What recompense is there for that? None.

Turning back to something you mentioned to me before we started this interview officially—someone asked you, why are you still here? I am turning this into, why do you do this work?

Amelia: I don't think that putting yourself in other people's shoes comes naturally to everyone. I don't know if they don't tap into it. We have to be plugged into this place and these people before we can understand what to do. We know the problems. You don't need a feasibility study for that.

To be able to do something with that, you have to be able to put yourself in someone else's shoes. We talk about bringing other people here, if we weren't tuned into what it is like here, if I were not tuned into what Jason's experience might be like here, as we are building this, because what we are suggesting is huge, it's not a band-aid, it's not a program, it's a paradigm shift but built on latent resources and things that worked here before but would work here again.

If we don't do this, now, then it won't be. Someone else will come and decide, using the resources here, the real estate here. It's cheaper here. Someone else will come in and they won't be tuned in to the people that are still here. The things we are proposing work with what is here and we are asking the community to do this with us. We want people here to have a stake.

What ties all this work together is also understanding health and having a healthy community. We can supplant the food purchased with food produced and employ people in this region. The people who do this work should be important and well-paid because they produce food not only for this region but potentially for regions beyond.

Crystal: Backing up, how did you get from 2015–2016 to here?

Amelia: Working in a department of health to form a coalition, and part of my job was to bring stakeholders to the table, and figure out what we could do make the region healthier. At the time, there was a big push for farmers markets. I went to a meeting in Tazewell, VA and met Jason and some folks from McDowell County, WV. Just like everyone else, I was surprised—there was farming in McDowell County?

We shared an emphasis on health, on healthy food, on providing food to the community. I started working with them almost as a proxy, because as producers they didn't have time to go to every meeting. We were working at that time on getting farmers markets operational around the region, or places in McDowell County. We were spinning our wheels trying to talk about food, nutrition, health, and how to connect these dots in the community.

They were taking what they grew to farmers markets throughout the region in WV and VA. They were doing this, but sitting there all day long, and no one was coming to the farmers markets. You had these beautiful structures in some communities for farmers markets and they were not being attended, they were not even being used. We did work at senior citizen centers because they get senior vouchers. We were trying to take away as many barriers as possible to connect people with this healthy food.

We began to see that this model does not work here. We are trying to fit into something that people in more populous areas are telling us is the way to get food into people's hands. We even tried doing a local CSA [community-supported agriculture—a local food delivery kind of service] in McDowell County for people in the Health Department and Department of Social Services. It took a while to understand that this was not going to work here.

Every single experience, from personal to my different titles at the health department to working with Jason and producers in McDowell to your research, shows me a heavy lift needs to be done, and this heavy lift is more in the mind. We have lots of resources here. The work is also in building community trust and building the support that we need from the powers to be to understand what we can do here. We don't despise small beginnings, but we tried the small beginnings. We have done all the things recommended like farmers markets.

We need major investment to have a major impact. We have research and historical data, and we are working to understand what the impact could be by working with some other organizations to do this. We have jumped through the other hoops. If you look around, the region is telling us what is possible here. You can look around and see maple, see bees, see what naturally occurs here, and it is telling us what to do.

You can look around everywhere else and see what you can be doing there. We are not suggesting that we should do this by ourselves. We are trying to connect to other areas, populous places, and create a resource exchange, how you build something that works. We can produce things here in high demand not just in urban places in the United States but around the world.

Crystal: Tell me more about that.

Amelia: The honey that we produce here is a superior product. We have had mentors help us understand how we can create a product here that people will want everywhere. As a beekeeper myself, this is another product produced here, and the environment, and trees, and so on work with this. We can add more pollinator zones to build up a demonstration facility to support an apiary. We understand that the honey we produce here is because of our nature here. The honey here rivals any honey I have had. And that is only one product.

We spoke with someone recently and were asked if we would want to learn more about programs planting trees to diversify hardwood—there is a multi-billion-dollar industry for this product. This is also the kind of thing that this region could support.

When we started we almost had to beg to get in front of people to tell them what we were working on. We are looking to change the identity of where we are. We want people to be proud of where they are from, and we want everyone else to be proud of this place, too.

Jason calls this the last frontier, a gold mine. Some see that as crazy as we are seen as the worst of the worst. It is the right place for change to happen. Some folks say we can only go up from here, but we could do worse. We should be showing people what it looks like for an area to get its act together. There is no better place to show what change can look like. We've talked about what if this place were a utopia for food production, diversity—recreating an area while staying true to what makes this area great. Again, if not us, then who? And if it is someone coming in, it would not be what you want.

Crystal: I went looking for a guide. I wanted to figure out how we do this. I wanted some advice. If you could give people some advice, what would you give them if they are sitting there and facing what we face.

Amelia: Something that has informed all my work is something like the book *The Speed of Trust*.[1] One of my strengths is in relationship building. We need social capital. We needed relationships. We needed to understand all the different stakeholders, and the people in power, and how to navigate all of that. We had to learn the rules and adapt as rules changed.

Leading this organization the thing most helpful to me has been forming and nurturing relationships. When people trust you, they will help you and they want to see you succeed. We constantly have to work on this and show ourselves to be people who show up and do the right thing. That is always not a popular position. We are not here to line our pockets. This is about all of us—the people who are here and the people who will come here. Trust is such a huge piece for me. It is not about money or notoriety. We have been true to what we see as the endgame.

There is no better place to do what we are doing. You would have to be crazy to not to want to see this succeed. What we are proposing is not some small thing. It is really rebuilding this area: families, communities . . . we want to affect other businesses, other organizations. We want to collaborate and make this change. It is going to be evident.

NOTE

1. C. S. M. R., *The Speed of Trust* (Free Press, 2006).

Chapter 5

Toward a National Sustainable Agricultural Strategy

In 2011, I moved to North Carolina and took up farming in the foothills below the Appalachians in a mini-mountain range (the Brushy Mountains or the Brushies). One hundred eight miles (give or take a few) northwest of this farm, my grandmother had moved with her family from a coal tent camp in McDowell County, WV in 1890 to a tenant farm in Bastian, WV. Her parents had moved to McDowell County, WV in 1885 from Wilkes County, NC for her father to work in the newly opened mines of the Pocahontas Coalfield. This was before towns or homes had been built for the coal miners there. My grandmother had been born in McDowell County when her family lived in that tent camp. After five years of this work and life, her father moved the family two counties south to Virginia and back to farming. My grandmother married a local, and she and my grandfather lived on and worked the property of a local country doctor, providing food also to his family in addition to producing goods to sell, principally tobacco and milk.

My mother, the youngest of ten children in that family, recalled that this farming had kept my grandmother so busy that some years my grandmother did not venture out of sight range of the home where they lived. Among many tasks, my mother grew up digging potatoes out of the rocky ground (something she loathed), hand milking sixteen cows, and harvesting tobacco. As a young adult, she escaped farming, preferring to work in the local textile mills, the work available to White women from mid- to late-twentieth century in an adjacent industry to energy .

My mother had this response to my taking up farming—*with all that education? Have you done plumb lost your mind?*

Although she lived the last years of her life also on this farm in North Carolina, I am not sure I ever convinced my mother that the farming she had done was not the same type and kind of farming my husband and I were doing.

As I have addressed in earlier chapters of *Big Rural*, despite the short history of coal production in the Pocahontas Coalfield, manufactured coal miner and family identity, public relations campaigns from mining companies such as "Friends of Coal," "Coal Keeps the Lights On," and "I Love Coal" and local and national media depiction of Appalachia as "coal country" assist in keeping local conceptualizations of self and work locked on that single sector production.[1]

Within the last ten years, I was sitting at an energy conference at a table with a mining safety Ph.D. and next to a presenter from Harvard. He and I were incredulous when she was surprised at a joke I had made—that mining was not exactly done any more like when my Grandpappy Cook worked with a hard hat, pickaxe, bucket, and a light. For lack of better understandable terminology, I countered to her that mining was now highly technical and that miners even may be likely to have earned a two-year degree to operate enormous heavy equipment.

With the agricultural sector in our country—like with this presenter from Harvard and coal mining, through media, children's books, advertising, and branding, most of us maintain an antiquated concept of agriculture as some blend of a Farmer Brown with his few happy cows and little garden and big green tractor and a myriad of talking animals—much less the hard work of my mother's youth on a smallholding production farm in Appalachia. Even those of us who know differently struggle to imagine agriculture accomplished in any way other than as a large technical system (LTS) with mega machines, aqueducts, chemical fertilizers, corporate patented seeds, massive warehouses, 24-hour trucking, centralized meat packing, hundreds of factories, thousands of groceries, line item Cooperative Extension, and abundant (yet usually invisible) migrant farmworkers—a current group in kind akin to my grandpappy a 100 years ago as a miner, in that they themselves function even now as harvesting and agricultural technology.[2]

As outlined earlier in *Big Rural*, agriculture has remained an adjacent economic sector in central Appalachia and only deteriorated in major significance with the push nationally post–World War II for federal support for large holding corporate Midwestern and Western agriculture.

REINTRODUCING LOCAL/REGIONAL AGRICULTURAL METIS

At the same time as I have been researching, analyzing, and theorizing the Pocahontas Coalfield as a coalfield labscape, an LTS, Big Rural, rural industrial space, I have been engaged in smallholding farming and learning the practices and theoretical landscape of sustainable agriculture. In 2016, to this I added working in collaboration with folks reintroducing production farming to McDowell County, WV in the Pocahontas Coalfield. Piloting, planning, advocating, and analyzing together as a volunteer collective under a nonprofit we co-founded—we found that the field models promoted by West Virginia Cooperative Extension and the West Virginia Department of Agriculture, and by the United States Department of Agriculture (USDA) for sustainable agricultural market models (farmers markets, farm to school, etc.) did not translate into the topographical or economic situation of the Pocahontas Coalfield; however, we also found that there were certain productions that suited mountain agriculture and that could be scaled toward fiscal, production, and environmental sustainability. The interview and white paper "interludes" in *Big Rural* expand on these.

The potential to scale agricultural production, even in the brownfield rural industrial environment of the Pocahontas Coalfield, has become apparent. Coal and minerals are not the only abundant resources possible there. More recently, retooled coal companies now "natural resource companies" have restarted aggressive timbering. Among other possible productions, in this abundant timber stand are many acers that produce syrup, and, many flowering trees perfect for supporting apiaries.

To demonstrate the potential of these mountain appropriate agricultural productions as scalable beyond cottage to a viable economic sector, I recognize that orchestrated funded large policy or strategy from the federal government for sustainable agriculture would be a gamechanger for the Pocahontas Coalfield, in Appalachia more broadly, for single sector rural spaces across the country, and, as a model of agriculture, influential to other places around the world.

This is not to say that even other economic sector developments are not possible in a resource rich place like the Pocahontas Coalfield. Yet, along with recognizing the economic potential of mountain appropriate scalable agricultural productions, food and agricultural production are central to community, and central to community repair.

With respect to the environmental, health, and economic degradation of the local population of people remaining in the Pocahontas Coalfield outlined in earlier chapters of *Big Rural*, scalable locally directed sustainable agricultural productions offer new paths forward.

To get there, the kinds of significant investments of federal, state, and private capital akin to those that created a Big Rural industrial space such as the Pocahontas Coalfield need to happen. This time, though, there is the benefit of hindsight—we can envision not only the benefits proposed by someone like Jedediah Hotchkiss in the nineteenth century of a prospering coalfield industrial and community environment for the two Virginias but also the detriments and the end games of technical automation. For new sectors can assert improved and sustainable relationships to financing, technology, infrastructure, land use, community support, resource utilization and control, education, and health outcomes. We can assert measures beyond GDP and jobs created or retained for community and individual well-being as enumerated by Atlas Charles in *interlude* 2.

In this chapter, I move from imagining a National Rural Strategy in *chapter* 4 to a practicable National Sustainable Agriculture Strategy. Again, as I stated in the previous chapter, I don't anticipate the adoption of these by federal agencies; nevertheless, I offer potentialities and push us all to consider interventions that may veer strongly away from current models.

This is not an attempt to go back to the small holding mid-twentieth century farms in Appalachia described by Jerry Bruce Thomas and like my mother grew up working, but a thought experiment based in practice about a way forward. As society's backbone, how we want or need our society to be is directly reflected in the kinds of systems of agriculture supporting it. We can't shift to a new "greener" society in terms of green energy without also shifting to "greener" agriculture and including practices concerning access, equity, and the democratic citizen.

Elsewhere in this volume, I have discussed the need to open the black box on an industry and a sector toward unveiling their inner workings for consideration and review. I have critiqued the single sector economic space as stymieing and choking democracy and as spaces where private and work life collide. Many other theorists and researchers have described our food system in the United States leaving me no need to belabor their points in this volume. Despite discussions and corporations entering organic agricultural production, sustainable *community-centered and sustaining* farming orchestrated to feed locales or regions remains marginal.[3]

As I originally conceived this chapter, I drilled down on many aspects of our current conceptualizations and practices of agriculture as off the mark and as fantasy spun by media and politicians—from who farms to our relationship to death in agriculture to how we eat. After sitting with this, I realized that each of these topics has been well-covered elsewhere, and, still, could have volumes written about them before we settle on how to approach this massive system.

By contrast, what we need are potential plans—the will and permissions to reimagine. For the National Rural Strategy earlier in this volume, I posited oughts—what we need to do written in large strokes. In this chapter, I attempt to tie together what should be with paths—a strategy with direction and action.

BIG RURAL LTS AGRICULTURE AS STATE OF EXCEPTION—THE WHOLE WORLD AS LINDYTOWN, WV

Industrial scale agricultural production or "the agricultural treadmill" outlined by a set of reports from the United Nations and World Bank, surmises our current situation:

> On behalf of the United Nations and the World Bank, in a four-year-process, more than 400 scientists summarised the state of global agriculture, its history and its future. The outcome was the International Assessment of Agricultural Knowledge, Science and Technology for Development (IAASTD). The findings are uncomfortable and alarming: providing a warning on the misleading ways of the past and showing new ways forward.[4]

The associated research briefs paint a bleak picture.[5] Our current Big Rural agricultural LTS produces food efficiently, distributes this food quickly, all while also promoting food indentureship—that we become indentured to companies outside our region for feeding us.

We outsource our sustenance to companies that measure profits in billions such as Archer Daniels Midland, Cargill, Bayer, John Deere, CHS, and the Mosaic Company. In this indentureship, our direct communities lose the knowledge, the infrastructure, and the land for production. In this Big Rural, we lose civic and local control of the means of sustaining our lives, our tangible, physical selves. In addition, in Big Rural, we outsource our health, our wellness, our well-being to external profit-driven companies.

The answer to my mother's query I quote earlier in this chapter about my moving to a farm after I received "all that education" resides in my and my husband's return to practicing what we preach, to choosing to reengage, as we can, with life-sustaining metis. We understand it also as a privilege to devote any of our time to reskilling. Yet this reskilling ought not to be relegated to a personal choice but included as state strategy and policy.

Rather than GDP and measures of agricultural export, our state agricultural agencies ought to focus on the population in each state being able to be fed by farmers in that state, or at least farmers in that region, the kind of cross-statutory planning I describe as related to bioregions rather that our manufactured state borders.[6]

At the minimum, for resource security, we ought to have local/regional energy *and* local/regional food metis and production. To this, we can develop an iterative ethic of other measures of community and economy, of economic well-being, including the measuring of tangible, physical life-sustaining work, systems, infrastructure, environment, and dignity.

The externalized costs of Big Rural industrial agricultural production are large: gratuitous waste, gratuitous animal death, early human worker death, child labor, cheap food leading us to early graves, soil destruction, deforestation, methane gas released from food in the dump, fossil fuels burnt to transport food, packaging waste, phosphorus depletion, marine life choking on plastic cups and food trays, methane contribution to ocean death and warming.[7]

The world's largest societies have excelled in food production, but cheap food wrecked their environments and also contributed to their downfall.[8] The quality of the food we eat, where and how it is grown, the Big Rural industrial modes of farming we currently practice, and the severe detriment to the soil and climate make our LTS agricultural system a failure.

A recent *National Geographic* article on feeding the 9 billion people on our planet succinctly summarizes much of this Big Rural agricultural catastrophe:

> Agriculture is among the greatest contributors to global warming, emitting more greenhouse gases than all our cars, trucks, trains, and airplanes combined—largely from methane released by cattle and rice farms, nitrous oxide from fertilized fields, and carbon dioxide from the cutting of rain forests to grow crops or raise livestock. Farming is the thirstiest user of our precious water supplies and a major polluter, as runoff from fertilizers and manure disrupts fragile lakes, rivers, and coastal ecosystems across the globe. Agriculture also accelerates the loss of biodiversity. As we've cleared areas of grassland and forest for farms, we've lost crucial habitat, making agriculture a major driver of wildlife extinction.[9]

No "lifestyle" (vegetarian, paleo, vegan, pescaterian, etc.) eating choice remains unscathed, and no industrial scale practice, whether producing flora (nitrous oxide, runoff, biodiversity loss, clearing grasslands and forests) or fauna (methane, runoff, biodiversity loss, clearing grasslands and forests), remains innocent and both produce the death of climate change.

The death through climate change puts the world in the Achille Mbembe necropower state of exception I discuss in my theorization of Lindytown in Boone County, WV. The world becomes a Lindytown, and collectively, we are the Richmonds, aware, peering out at the large rock that is being shaken loose by corporate-powered technology and priority to destroy us.

As residents of that house, we are robbed of our democratic citizenship. In this scenario, corporations and a colluding state have usurped our democratic agency. We live in what is touted as a democracy without power over our lives, or, our deaths, as life-giving life-sustaining choices are usurped, tabled, and removed from possibility; we vote not on what will save us but about lesser options orchestrated so that we cannot loosen necropower's hold.

At least the Richmonds had the benefit of being cognizant of the threat to their existence. If democratic policies that address climate change are not front and center of all policy, then the choices we are given in our democracies are false and mask the doom impending. To be given choices about other items in the starkness of doom comes with no agency at all: it's like asking "tea or coffee" to someone in the death chamber; the real issues are circumvented.

If we stay as we are where we are, we inflict no price on the corporate imperative. The corporations continue doing what they are doing until any life-sustaining scrap is dead. Whether we live or die is insignificant to the corporation and its power. And, as I have emphasized elsewhere, "corporations" consist of people, scientists and engineers, and, people are the ones making these decisions on the collective fate of our planet as inhabitable by living organisms, whether we (the collective we of life) all die like mayflies in an acid mine drainage-infected stream.

From this starting point of corporate agricultural necropower of what feeds us and also killing us and the planet, I pivot to utilizing some of the practices of approaches and strategies I proposed earlier in *Big Rural* for the rural space and apply them to how else agriculture could be practiced, and, what a strategy for this could contain.

These strategies move us out of necropower to revitalize the democratic citizen, for now, and, for future generations. In necropower and a deathworld, we and subsequent generations lose our democratic citizenship; the large LTS-promoting corporations and the colluding State rob us of choice, of agency.

Farming *is* the backbone of society. Someone or something else producing food (and energy) frees the rest of us to follow other pursuits rather than spend most of our time and initiative producing or finding food. That production does not have to be relocated out of a community or region.

No one will rescue for us our democratic agency—not Calvary, and not the Cavalry.

Democracy is a practice, not only a set of theories. What would a freer more life-sustaining agriculture be? How do we reclaim metis in agriculture, and, support ourselves and our communities with sustainable production? What strategies in agriculture may make more Americans freer and steer us out of our current Big Rural agricultural deathworld?

NATIONAL SUSTAINABLE AGRICULTURE STRATEGY

Elsewhere in *Big Rural*, I discuss the importance of the strategy document in U.S. administrations and as directives for action. I then frame some potential directives for a National Rural Strategy toward reformulating State and other relationships with rural spaces in the United States to make the rural a space of more equity, more justice, more democracy, more sustainability. I cover ground such as:

- Decentralize LTS and Support Metis-Friendly Institutions
- Support Science and Engineering that Add More Knowledge rather than Flatten Metis
- Promote Science and Technology that Support and Steward the Rural
- Build on Regional Groups and Their Emerging Coordination
- Redefine Regional Success
- Stop Educating Rural Citizens to Leave the Rural.

In this same vein, in this current chapter, I propose recommendations for a National Sustainable Agricultural Strategy not as absolutes, but as springboards.

NATIONAL SUSTAINABLE AGRICULTURAL STRATEGY

Support and Promote the "Soil Path."

In the early 1970s, physicist Amory Lovins outlined a clear and absolutely necessary path to renewable energy.[10] Lovins called then for a veer from fossil fuel dependence to a "soft energy path" of renewables, and his call then only recently has begun to be heeded.

We must redesign agriculture to follow a "soil path" if we are to combat climate change and continue, for the most part, to feed ourselves and our communities and not leave our children and grandchildren, if they survive this century, living inside a cave, eating the insects they can catch as the potential outcome of our current Big Rural LTS paths as envisioned by William T. Vollmann in his well-researched *Carbon Ideologies*.[11]

The soil path makes rebuilding the U.S.' thinned soils, retaining soil, creating healthy soil, remediating soil, a top priority, for our planet and for carbon sequestration, for ensuring the potential for communities to feed themselves locally, and for nutrition. The soil path reverses our trajectory of death through agriculture.[12] The soil path encapsulates this question: does this (practice, policy, activity, proposed activity, industry) save soil or deplete it?

Recognize Soil and Farmland as Finite Resources. Develop Policies and Practices for Their Further Protection and as Necessary Resources Now and for Future Generations.

A couple of simple questions serve as a guide: Before any construction, new parking lot, other use, and so on—would this ruin land that has been or could be used for growing? If so, how do we accomplish XYZ and still steward this soil for growing food for our own and subsequent generations? Add to this, how can we bring back to life soil that has been decimated?

For example, a few years back while living in Blacksburg, VA, I watched former dairy and cattle farms, places of beautiful chocolate cake-looking soil and mature pastures, get covered by cheap housing developments. That soil and that grass took generations to build and now are lost, maybe, forever, to housing, in a region where soils have already become marginal due to mono-farming and timbering. Soil is a finite resource, our essential resource, and an ignored resource. We continue to abuse what sustains life. Eroding soil is not just washing away. Eroding soil reduces in capacity to support human (or complex) life. Just as we consider safe-guarding other aspects of existence for future generations and use, soil also must become precious, not regarded as just "dirt." Safeguarding soil includes recognizing its essential role, and, reworking economic and government systems to uphold and protect soil. What if soil protection guided economic and agricultural development?[13]

Rethink the Goals of the Farm Bill.

Make soil health, soil retention, and planning for food production co-located with the communities the food feeds the main agricultural production priorities for the Farm Bill. Cut out corporate or corporate-backed lobbyists from its creation or recommendation—follow the money trails on any nonprofit advocacy as well. Beyond the Farm Bill, create a National Rural Policy/Strategy. In the United States, the Farm Bill (for a range of current and past political will and corporate money) serves as the catch-all for the natural resource push–pull of rural economic development, regardless of the dominating economic sector in a specific rural location (mining, tourism, and agriculture). Thus, much rural economic development, regardless of local industry, falls under the U.S. Department of Agriculture.[14] This is absurd, but it is fixable.

Decentralize Food Production: Disperse Growing and Production for National Food Security.

Food production must be dispersed throughout the United States for increased food security and community health. This is essential to the breaking up of Big Rural.

The centralization of most of our food production as an LTS in the United States into only a handful of states over the last 100 years, with California and Iowa nearly producing double that of the next three states combined, makes the United States and our individual cities and communities less secure.[15]

Mass disruption in our food production and distribution system does not require an invasion, but something like reducing food safety confidence at best, or staged disruptions of highways essential to trucking distribution of food across the United States, like Highway 58 out of Bakersfield,

CA, or highways through Southern California and/or Highway 80 out of Sacramento.[16]

Beyond pandemics, the potential culprits for food distribution disruption pile up in terms not only of what, but of when: climate, fires, water, the Colorado River, hurricanes, economic breakdown, corporate neglect, transportation stoppage, telecommunications, energy or nuclear breakdown, political turmoil, much less any kind of outside or foreign intervention such as jamming nonmilitary operations or the food crises caused by Russia's invasion of Ukraine.[17]

For example, as of the writing of this book, one U.S. farm feeds 165 people annually here and abroad (through exports). The global population is expected to increase to 9.7 billion by 2050, which means the world's farmers will have to grow about 70 percent more food than what is now produced.

Diverse production for local or regional potential sustenance and sustainability across the United States ought to be a major focus of national security in our age of climate change.[18]

Access to Local Land.

Communities, cities, towns, counties, citizens, nonprofits, and universities, ought to assess local land amenable for growing or raising food, then, consider plans to steward growing on and stewardship of that land. This ought to be part of every community's planning for community viability and security.

Each U.S. community needs farmers/producers, significant food production infrastructure, investment, and marriage between soft diversified energy production and soil path agriculture in an implemented plan.

Revise University-Based Extension to Feed Communities.

Extension was created to distribute scientific practices to farmers.[19] Science as a practice or performance is not a monolith: soil science is science; likewise, Monsanto seed proprietary practices are also science-based. Thus, Extension ought to focus on the specifics of necessary farm stewardship: soil health, water health, pasture health, farmer health and well-being, farmworker health and well-being, which all contribute to community health both structurally and physically by supporting the people and land to grow more nutrient dense less carbon intense food.[20]

Universities pay attention when significant federal Research and Development (R&D) money is invested in an issue.[21] I had as much said to me by the heads of R&D at Virginia Tech in 2016—if there is no request for proposals from the federal government, then XYZ must not be a problem.

Highlight and fund soil retention and health as well as soil reserve as major national priorities, especially under the Department of Defense (DOD), as a means of better securing our country's resources and food supply and assuring civic harmony.

In Extension and other agricultural training and education also:

- Train to support soil health and rebuild thinned soils.
- Diversify environment and geographic agriculture and agricultural education.
- Adapt practices and science to the potentials of location and region.
- Research and design physical infrastructure to be made available to producers and entrepreneurs.
- Analyze how to achieve a reliable labor force for management when operations scale.

Grow Agricultural Infrastructure.

Grow agricultural infrastructure alongside growing producers and the other essential jobs/businesses related to production such as logistics, supplies, and support services. Create paths and practices for new producers in your region and all the accompanying jobs of the adjacent needed infrastructure.

The tremendous myopia of the sustainable agricultural movement has been its focus on growing production without also focusing on growing infrastructure and reducing risk for sustainable farmers. In 2014, my husband and I totaled up seventy support services and players within an hour of us that allow us to farm, from custom hay operators to heavy equipment rentals to Southern States stores to pasture seed suppliers to Tractor Supply stores to diesel mechanics to FedEx branches to accountants familiar with agricultural laws to insurance agents that can and know how to ensure a farm to banks that understand farm cash flow.

Production is only one facet of many of farming. Growing essential infrastructure and adjacent businesses and services must happen alongside growing producers. Otherwise, your region or community will never really produce much as entry point producers cannot shoulder the risk of operating a business in a location that does not offer the infrastructure and support services producers need to be able to thrive.

In our region, the farmers market is not a serious model for production due to not having the urban consumer markets locally or being within reasonable distance of large cities for consistent farmers market sales to earn a living wage or pay for the costs of production. Plus, most food most people in the United States get does not come from farmers markets. The latter is marginal in both market share and consumer access, at best.[22]

It is unreasonable to expect someone who has identified the production of food as a career goal also to be an expert in marketing, sales, consumer education, pricing, branding, and so forth as is expected in the farmers market model.

Thus, these additional priorities under infrastructure extend the imperative to grow agricultural infrastructure. They are not in order of importance.

Invest in year-round marketplaces and sales for producers in your region, and location. A series of farmers focus groups I did in 2014 in our region sparked discussion as to why consumers go to Whole Foods rather than the farmers' market: at the former you can buy tomatoes AND buy wine, shoes, meat, dishwashing liquid, and dessert there. Farmers and consumers have little time, and sales and access must be easy. Focus on building on current locations and businesses for access. Reduce the risk for current business owners to sell locally. Grocery store produce managers run the risk of losing lucrative delivery by big players, like those from California, if they forego those for smaller producers for any or part of the year.[23] Find ways to aggregate to ensure volume. Plan regional production and support regional production and paths to sales. And, create infrastructure to make local ready to eat food available.

Refocus the ethic from strict definitions of "local" to supporting soil stewarding farms, people, and practices. The son of a McDowell County farmer was in significant high tunnel vegetable production with friends. The best farmers' market along that parallel is an hour and a half away in Blacksburg, VA. Though producers of color and youth and though stewarding soil health in a coalfield region, they were deemed "too far" to sell at the most lucrative farmers market in the region. In my view, an entire book could be written about this dividing line and its short-sighted ethic. Consumers want and expect certain produce and products year-round. During your immediate locale off-season, if you can support a soil stewarding producer 300 or more miles away, then why not do it, to extend both local purchase options for consumers and that farm's sales?

Steward unlikely as well as the "bird in the hand" candidates and unlikely places into a range of productions. Support the farmers we have: As farm employment has plummeted worldwide, make creating farmers/producers for every community a priority—you have your family doctor, your family dentist, your teachers—then also have your family farmers, your local producers, your local beekeeper, your local orchardists, and so on.

Staff infrastructure for local production. Don't overpromise jobs: Though the renewable energy sector promises jobs, no large industry is in the long-term business of hiring people, this includes the soft energy (renewables) and the soil path. Most large industry avoids hiring if they can. As I covered in other chapters automation remains a threat to the U.S. workforce. The jobs in renewable energy come mostly in setup, which requires flexibility and travel.[24]

As a 2016 Obama administration report demonstrated, if you make less than $20 an hour, your job will probably be automated away in your lifetime.[25] Thus, focus on creating work in agriculture that satisfies community security and income but that cannot be automated away.

The most obvious bird in the hand remains stewarding and supporting the U.S.'s 2M+ migrant farm workers to become farmers. Farming is difficult, and you need people who understand the imperatives and realities of this hard work. We must create a path for significant amounts of farmworkers to become farmers.

Assist current farmers with land stewardship, legacy planning, and aggressive matching of future farmers to this land. Assess what essential infrastructure and services your community needs to break free of food indentureship: When planning your community for food security and away from food indentureship, assess together what knowledge becomes valuable. How and who do you have in the hand (assets) and who and what do you need? Who suddenly becomes more valuable? How do you attract those citizens now?

Support agricultural learning in every school. Install agricultural and production learning and practicum paths into every public school, community college, and university with an adaptation for production in that school's current environment, whether urban, mountain, rural, beach, fishing village, or suburban lawns.

Make zero waste from food waste. Eliminate food waste by channeling it into repurposed production of feedstock for livestock and into compost. Per person in the United States, we waste over a pound of food per day. Investment must made in every community to create facilities that convert institutional preserved food waste from grocery stores, colleges, universities, restaurants, and schools into systems for conversion to livestock feed or into compost.

Food coming into your location is importing fertility into your local system. For example, China currently sources milk from California, and what it is actually importing is California's soil fertility.[26] Thus, by capturing food and food waste sourced from outside your region, you are increasing the fertility and feedstock in your own region. Livestock producers across the country are currently reliant upon grain imports to feed their stock. This conversion of waste to feed makes a local protein boosting source possible without over-reliance on sourced grains.

Federal USDA, National Science Foundation, and Defense research and development grants ought to focus on efficient systems for individual consumers and apartment dwellers to also be able to contribute to compost creation and aggregation.[27]

Require all food packaging not recyclable to be biodegradable. Though plastic is an amazing substance in terms of its flexibility and use on the farm in everything from electric fencing to feed buckets, plastic is now entering our bodies and those of animals, contributing to a range of potential maladies.

We must reduce our dependence on plastic as a food transport and packaging medium.[28]

Prepare to have fewer pets. In the United States, we need to drastically reduce the number of stray and unwanted pets to reduce agricultural resources (soy and livestock) for pet food. Research points to this for reducing greenhouse gases from a reduction in pet food consumption alone, not taking into consideration the environmental impacts from other consumer goods produced for pets.

Pets have a large environmental impact, especially in their meat and soy consumption. If we commit to less meat and soy production in the future, that will also mean feeding fewer pets.

In an estimate I compiled, per year we in the United States feed our 90 million dogs the equivalent of 67.5 million cows at three-fourth pounds of meat per day, or, the equivalent of 73 chickens each; U.S. cats, 95.6 million, eating a one-fourth pound of food a day, comes to about 17 million cows, or, 19 chickens—thus, we kill many animals to feed one companion animal. Additionally, millions of wildlife animals, especially songbirds, are killed each year by cats left or kept outdoors.[29,30]

A growing international research literature exposes these environmental issues of pets. Gregory S. Okin adds mathematical accuracy beyond my casual calculation above. Okin estimates that "dog and cat animal product consumption" puts 64 ± 16 million tons of CO_2–equivalent methane and nitrous oxide into the air annually. He suggests that having other kinds of pets not dependent on the consumption of other animals may be one path, as well as focusing on alternative sources of protein for pet food.[31] Some debate exists on methods of calculation, but overall, pet food consumption and its modification is considered a potentially solvable issue in terms of reducing carbon footprint.[32] Spay and neuter programs are like vaccines for the environment and for the future. Fund them to be free or nearly so just like human vaccine programs.

Make taxpayer funded, federal money receiving public universities focus first on serving the public interest. Taxpayer funded colleges and universities ought to focus on educating their attendees to solve community and public problems or collectively address climate change, policy, infrastructure, or even local problems in their host communities such as urban food production, food deserts, unemployment due to automation, or brownfield rehabilitation. Fund them for this research directly and block them from receiving corporate research grants, or endowments with strings attached.

Just as the organization Public Citizen focuses on getting big money out of politics, a similar movement and set of federal policies need to happen with getting big money out of land grants and public research universities.

Mission drift has occurred across the board with public research universities—there has been an explosion in corporate-sponsored centers and research and administration, all the while expanding the role of adjuncts and part-timers to teach. This diminishes the education for attendees and causes universities to compete for corporate funding, rather than focus on turning out the best problem solvers possible for the greater public good.[33]

What may seem a huge windfall grant to a university to do research for a corporation results in mega-savings for the corporation, as they do not have to source all of the infrastructure, the graduate students, the sites, the talent, the adjacent facilities, and equipment to conduct their own R & D.[34]

Public universities ignore their local communities and more global problems in favor of chasing corporate endowments and federal money. One of the additional reasons for the disconnect between universities and their surroundings also stems from the continental U.S. model of "going away" to college, meaning that students do not develop their critical thinking skills in the context of their own communities, but in one where they reside for a finite time, and exist in a bubble separate from the surrounding population and place. I contrast this to popular student movements in other countries where students and local communities engage in mutual support and solidarity. Even student protests over the past decade with community support in the U.S. colony of Puerto Rico reflect an alternative possible dynamic, with large community support for student demands, as university students are from the community proper, or nearby.[35] We don't expect or allow corporate funding of our public elementary and high schools . . . why are they allowed in higher education?

The best scientific minds at our taxpayer funded public universities ought to be focused on science and technology for the public good and basic science for expanding scientific knowledge. Let corporations fund their own R & D and get them off taxpayer subsidies at public universities, which leads to the last suggested strategy.

Federal research agendas must focus on funding science and technology that serves the public interest or at least supports basic science for expanding scientific knowledge. The public interest can be defined by adhering to the commons and the Constitution, supporting policies that increase or at least support democracy, and engaging in ethical decision-making. *Big Rural* implicates scientists and engineers in creating the spaces we inhabit, and, explicates their role in enabling modern society.[36] Scientists and engineers are part of our society and in our modern context, their research circumscribes our economic and physical potentials.[37]

The field of the Science and Technology Studies of Society approaches scientists and engineers as subjects themselves with cultures, practices, ethics, history, structures, and opinions, worthy of study just as an anthropologist may study a village culture or a sociologist's social movements. We can investigate and work together for cultural shifts among scientists and engineers and the purpose of their production.[38]

Sustainable Agriculture Infrastructure = The Milestone to Sector Development

In this present chapter, I have introduced a National Sustainable Agriculture Strategy that increases democracy by breaking up Big Rural agricultural LTS. Not only theoretical or drawn from academic research, this proposal is also rooted in on-the-ground work.

The current federal situation for sustainable agriculture runs largely through the USDA and comes in the form of producer or nonprofit grants, usually in the hundreds of thousands of dollars, maybe in a couple of million, very rarely in the levels or grants available from the DOD or Energy for their priorities. The federal government, as I pointed out with the jump in renewable investment under the Obama Administration, invests tens of millions, if not billions, to shift or to support sector development. In terms of federal support for sustainable agriculture—while getting $250,000 to a food hub is great, getting $250 million into the region or $2.5 billion indicates the level the feds play on to signal they are serious. Sustainable agriculture currently competes for this cottage-industry level of $250,000. Sector development activity will remain cottage until real investment becomes justified through science and economic research and then funded to sector levels by the federal government. No one will imagine these other sectors in the rural if we don't. No one will lobby for this kind of investment in the rural and elsewhere if we don't.

Not only for our subregion but across the United States, sustainable agricultural parks ought to be supported to reignite local and regional sustainable agricultural production and provide more food and thereby more civic security, and, less food indentureship. Regions ought to work together to produce regionally appropriate sustainable agricultural plans, drilling down to the level and analysis of necessary infrastructure, science, technology, education, policies, support industries, and so forth.

Imagine the game change of a zero-waste sustainable agricultural park open for use to a region's producers. This facility could serve as a central hub for sustainable agricultural and woodland harvest development expanding outward with honey processing, tree sap processing, herbal and medicinal harvest, a food waste conversion facility, a distillery, a brewery, a fermentation facility, a food processing facility, a dehydration facility, a cooperatively owned meat processing facility and kill floor, a micro-malter, storage, cold storage, aggregated services for producers, and so on. This space should include facilities and programs for business cross-pollination, community participation and education, engagement with renewable energy and sector development, and technical and innovation support. These parks and the products coming from them ought to be sustainable in sourcing and through earned revenue, rather than long-term grants or subsidies.

As I stated in chapter 4 with respect to the draft of a National Rural Strategy, any strategy is only as good as the institutions enacting it. Research agendas can inhibit or promote liberty and equity. I proposed here in chapter 5 specific strategies toward steering us out of the Big Rural democracy deficit, single sector economic work-civic totalitarianism, and the flattening of metis by scientific and technical practice and machines. While federal investment is important, the social change sector cannot wait on or expect it before piloting, planning, or advocating for what is possible.

NOTES

1. Prianka Srinivasan, "Appalachia's fickle friend." *Jacobin*. December 5, 2017, Accessed July 26, 2023, from https://jacobin.com/2017/12/appalachia-friends-of-coal-mining-industry-jobs-environmental-just-transition.

2. Review statistics at: "Farm Labor," Economic Research Service, U.S. Department of Agriculture, updated March 22, 2023, https://www.ers.usda.gov/topics/farm-economy/farm-labor/

3. Margriet Caswell, Keith Fuglie, Cassandra Ingram, and Catherine Kascak, *Adoption of Agricultural Production Practices: Lessons Learned from the US Department of Agriculture Area Studies Project. Agricultural Economic Report No. AER792*. Economic Research Service, U.S. Department of Agriculture, January 2001, https://www.ers.usda.gov/publications/pub-details/?pubid=41202.

4. "About the IAASTD Report." Weltagrarbericht. Accessed August 11, 2023. https://www.globalagriculture.org/report-topics/about-the-iaastd-report.html.

5. Access the research briefs through https://www.globalagriculture.org/.

6. Note that the success of agriculture is measured in GDP: https://www.ers.usda.gov/faqs/

7. Borja G. Reguero, Iñigo J. Losada, Fernando J. Méndez, "A Recent Increase in Global Wave Power as a Consequence of Oceanic Warming," *Nature Communications* 10, no. 1 (2019), https://doi.org/10.1038/s41467-018-08066-0

8. Laura Parker, *Parched: A New Dust Bowl Forms in the Heartland*, National Geographic Food, May 3, 2021, Retrieved December 11, 2022, from https://www.nationalgeographic.com/science/article/140516-dust-bowl-drought-oklahoma-panhandle-food. See also Dana Coulter, "The Fall of the Maya: 'They Did It to Themselves.'" *NASA Science*. October 6, 2009. https://science.nasa.gov/science-news/science-at-nasa/2009/06oct_maya.

9. Jonathan Foley, *A Five-Step Plan to feed the World.* Feeding 9 Billion - National Geographic. 2015. Retrieved December 11, 2022, from https://www.nationalgeographic.com/foodfeatures/feeding-9-billion/

10. John Tierney, "A Gift from the '70s: Energy Lessons," *The New York Times*, October 7, 2008. https://www.nytimes.com/2008/10/07/science/07tier.html; Amory B. Lovins. "Non-Nuclear Energy Strategies." *International Relations* 5 no. 3 (April 1976): 998–1010, https://doi.org/10.1177/004711787600500302.

11. Nathaniel Rich, "The Most Honest Book about Climate Change Yet," review of *Carbon Ideologies* by William T. Vollmann, *The Atlantic*. Atlantic Media Company (September 9, 2018). https://www.theatlantic.com/magazine/archive/2018/10/william-vollmann-carbon-ideologies/568309/.

12. A. Edward Johnston, Paul R. Poulton, and Kevin Coleman, "Chapter 1 Soil Organic Matter: Its Importance in Sustainable Agriculture and Carbon Dioxide Fluxes," in *Advances in Agronomy*, Vol. 101 (2009): 1–57.

13. Peter M. Kopittke, Neal W. Menzies, Peng Wang, Brigid A. McKenna, and Enzo Lombi, "Soil and the intensification of Agriculture for Global Food Security," *Environment International* 132 (November 2019): 105078, https://doi.org/10.1016/j.envint.2019.105078

14. See Congressional Research Service, *In Focus*: *Farm Bill Primer*: *Rural Development Title*, by Lisa S. Benson for Congress, February 9, 2022, https://www.everycrsreport.com/files/2022-02-09_IF12038_f7884223f4b9ed8d98ee4d5a0b0f8945226971a9.pdf. For a primer on Farm Bills, including the rural development's place under Title VI, see Congressional Research Service, *In Focus*: *Farm Bill Primer*: *What is the Farm Bill?*, updated June 16, 2023, by Renee Johnson and Jim Monke, IF12047, https://crsreports.congress.gov/product/pdf/IF/IF12047

15. Rob Cook, "Rankings of states that produced the most food," *Beef2Live*, updated July 24, 2023, https://beef2live.com/story-states-produce-food-value-0-107252.

16. "Outbreak of E. Coli Infections Linked to Romaine Lettuce." Centers for Disease Control and Prevention. Centers for Disease Control and Prevention, January 9, 2019. https://www.cdc.gov/ecoli/2018/o157h7-11-18/index.html; Mark Bittman, "Everyone eats there," *New York Times Magazine*, October 14, 2012, https://www.nytimes.com/2012/10/14/magazine/californias-central-valley-land-of-a-billion-vegetables.html.

17. Kati Pohjanpalo, "Finland Investigating Russia for GPS Jamming: 'Civilian Flights Have Been Put in Danger," *Bloomberg News*, November 18, 2018, https://nationalpost.com/news/world/finland-investigating-russia-for-gps-jamming-civilian-flights-have-been-put-in-danger; Tarek Ben Hassen and Hamid El Bilali. "Impacts of the Russia-Ukraine War on Global Food Security: Towards More Sustainable and Resilient Food Systems?" *Foods* 11, no. 15 (2022): 2301, https://doi.org/10.3390/foods11152301.

18. Christine Whitt, Jessica E. Todd, and James M. MacDonald. *America's Diverse Family Farms*: *2020 Edition*. Economic Information Bulletin No. (EIB-220), U.S. Department of Agriculture, Economic Research Service, December 2020. https://www.ers.usda.gov/publications/pub-details/?pubid=100011.

19. Everett M. Rogers, "The Intellectual Foundation and History of the Agricultural Extension Model," *Science Communication* 9, no. 4 (1988): 492–510, https://doi.org/10.1177/016402598800900 4003

20. David R. Montgomery, Anne Biklé, Ray Archuleta, Paul Brown, and Jazmin Jordan, "Soil Health and Nutrient Density: Preliminary Comparison of Regenerative and Conventional Farming," *PeerJ* 10 (January 27, 2022): e12848, https://doi.org/10.7717/peerj.12848

21. Matthew Rafferty, "The Bayh–Dole Act and University Research and Development," *Research Policy* 37, no. 1 (2008): 29–40, https://doi.org/10.1016/j.respol.2007.06.010

22. Garry Stephenson, Larry Lev, and Linda Brewer, "'I'm Getting Desperate,': What We Know about Farmers' Markets That Fail." *Renewable Agriculture and Food Systems* 23, no. 3 (2008): 188–99, https://doi.org/10.1017/S1742170507002153.

23. Respondents, Sustainable Agriculture Priorities Survey, 2014, Cook Marshall.

24. Michael Renner, Celia Garcia-Banos, and Arslan Khalid, *Renewable Energy and Jobs: Annual Review 2022*, IRENA. https://policycommons.net/artifacts/2685048/renewable-energy-and-jobs/.

25. Kristin Lee, "Artificial Intelligence, Automation, and the Economy," *What's Happening* (blog), National Archives and Records Administration, December 20, 2016, Retrieved July 29, 2023, from https://obamawhitehouse.archives.gov/blog/2016/12/20/artificial-intelligence-automation-and-economy. This blog announced the release of the Administration's document by the same name.

26. Lynne Terry, "While small dairy farms close, this mega-dairy is shipping milk to China." *The World from PRX*. December 5, 2018, accessed December 11, 2022, from https://theworld.org/stories/2018-12-05/while-small-dairy-farms-close-mega-dairy-shipping-milk-china.

See the USDA "endorsed" Chinese trade show SIAL China at https://www.sialchina.com/.

27. Krista L. Thyberg and David J. Tonjes, "Drivers of Food Waste and Their Implications for Sustainable Policy Development," *Resources, Conservation and Recycling* 106 (January 2016): 110–123, https://doi.org/10.1016/j.resconrec.2015.11.016

28. Richard C. Thompson, Charles J. Moore, Frederick S vom Saal, and Shanna H. Swan, "Plastics, the Environment and Human Health: Current Consensus and Future Trends," *Philosophical Transactions of the Royal Society of London, Series B, Biological Sciences* 364, no. 1526 (July 27, 2009): 2153–66. doi: 10.1098/rstb.2009.0053. PMID: 19528062; PMCID: PMC2873021.

29. Gregory S. Okin, "Environmental Impacts of Food Consumption by Dogs and Cats," *PLOS ONE* 12, no. 8 (August 2, 2017), https://doi.org/10.1371/journal.pone.0181301

30. Rachel E. Gross, "The moral cost of cats," *Smithsonian.com*, September 20, 2016, Retrieved December 11, 2022, from https://www.smithsonianmag.com/science-nature/moral-cost-of-cats-180960505/

31. Okin, "Environmental Impacts of Food Consumption by Dogs and Cats."

32. Peter Alexander, Aiden Berri, Dominic Moran, David Reay, and Mark D.A. Rounsevell, "The Global Environmental Paw Print of Pet Food," *Global Environmental Change* 65 (November 2020): 102153, https://doi.org/10.1016/j.gloenvcha.2020.102153.

33. Michael Rustin, "The Neoliberal University and its Alternatives, *Soundings* no. 63 (July 2016): 147–76, https://doi.org/10.3898/136266216819377057

34. Heny Giroux, "Neoliberalism, corporate culture, and the promise of Higher Education: The University as a Democratic Public Sphere," *Harvard Educational Review* 72, no. 4 (December 1, 2002): 425–64. https://doi.org/10.17763/haer.72.4.0515nr62324n71p1

35. Puerto Rico has seen waves of student protests, including in the past decade, around tuition hikes and austerity measures.

36. Corinna Porteri, "Sergio Sismondo: An Introduction to Science and Technology Studies," *Theoretical Medicine and Bioethics* 33 (October 1, 2011): 233–35, https://doi.org/10.1007/s11017-011-9197-6

37. Stephen M. King, Bradley S. Chilton, and Gary E. Roberts, "Reflections on Defining the Public Interest," *Administration & Society* 41, no. 8 (November 6, 2009): 954–78, https://doi.org/10.1177/0095399709349910.

38. Mark B. Brown, "Politicizing Science: Conceptions of Politics in Science and Technology Studies," *Social Studies of Science* 45, no. 1 (December 23, 2014): 3–30, https://doi.org/10.1177/0306312714556694.

Interlude 4

A White Paper as a Community Act

I cap the interludes in *Big Rural* with sections of a white paper that I principally authored, but which would not have been possible without the collective work I did with others in piloting how to proceed in the rural single sector industrial space of McDowell County, WV, to enact local change, and to understand what kind of larger policy or strategy would be needed to make true shifts in the circumstances current in the Pocahontas Coalfield. This white paper, as of the time of the interviews in the interludes (fall 2022), stood as a statement of the direction of our joint work in the nonprofit we cofounded—Economic Development Greater East (EDGE is still active but I have since shifted to other on-the-ground projects, see the Epilogue). In the white paper, we assert vision, direction, and calls to action. A white paper is a document of direction, of strategy, and not an academic paper up for peer review. Written two months before the interviews in the Interludes, this white paper elevates collective work, and including it in this volume, forges our common expression of the collective humanity and dignity we advance.

THE WHITE PAPER

In the late 1990s, I visited the newly built Jewish Museum in Berlin, Germany. The architect Daniel Libeskind designed the building so that you enter the museum through the lowest point in the history of Jews in Germany. The design engages directly with totalitarianism and its dismantling of our collective humanity. In this lowest point, the floor of the building tilts. A memorial disorients

From the lowest point in the building, the history, the engagement, the potential rises.

Hope.

Hope is found.

The rock had already fallen and destroyed all life below.

The path forward demanded dignity.

The path forward demanded containment.

The path upward demanded a new day.

I am not so much of a fool as to compare the Pocahontas Coalfield to the killing fields of National Socialism. I am not that ahistorical or arrogant.

Each historical event, if it can be analyzed as "an" event, stands in history on its own.

Even with the Pocahontas Coalfield as a type, there is no *historical* law for this type.

I argue in this volume for *Big Rural* as a useful conceptual tool.

To quote the tremendous author and politically engaged poet Pablo Neruda, whom I also quote in discussion in the Preface, and whose poem "Ode to Enchanted Light" (1973) I have had in the back of my mind while crafting the interludes. We are creating a world that is

a glass overflowing
with water.

This white paper points the way toward the light in Neruda's tree canopy. This white paper is a political act.

We have been led to believe we don't merit protest, much less poetry.

In their faces, we assemble facts, potential, possibilities.

We hope this same for you.

It's yours to do, and you are not alone.

EXCERPTS: OCTOBER 2022

With the climate and food, national, and civic security issues at stake in the United States, we at EDGE were clear in our intent when titling this white paper "Appalachia Has to be in the Game." To continue to discount or ignore the land, location, natural resource, climate, agricultural production, and water supply potential for the Eastern Mountains of the United States toward averting national food and security crises in the short-term, as well as those on the horizon, amounts to a crisis of political will, and even to continued regional health, economic, and civic peril.

[I would argue this is true of any single sector rural space—and I argue that elsewhere in this volume. Nothing *essential* predicates an area become single sector. Rust Belt. Corn Belt. Banana Republic. Coal field. Each region can and must be again allowed its latent diversity of environment, its redevelopment of human-scale and right size wealths and response.]

Appalachia ought to be front and center as the United States, from the United States Department of Agriculture to the United States Department of Defense, turns its attention to planning and enacting the civic and private response to the demands of our current century, and, as we look to preparation for the next. Food production in these Eastern Mountains must be a large part of our country's current and future plans.

Farming Is Much More than Flat Acreage and Massive Tractors

The scientization of American food production since the 1880s, with the spread of Agricultural Extension, has continued to favor environments that can be managed by large-scale machines and controlled through chemical and technical means to make the uncontrollable aspects of farming (weather, pests) less destructive. The by-product of this approach is the elimination ofa lot of the complex knowledge of land and environment, a flattening of natural complexity that also can result in fewer pests and in protection from weather.

Along the way, understanding and producing in a mountain environment, along with scaling to a reasonable size in a mountain environment, has become lost knowledge, and, under-investigated.

As a nation, we have been groomed since the mid-twentieth century through media and advertising to have fixed ideas about where and who can produce food, when, by contrast, for most of the U.S.' existence, most Americans of most backgrounds produced food for family, community, and surplus consumption.

Mountain farming can be renewed through new trends in marketing, sales, safety, production models, product diversification, and client relations. It deserves significant research allocation, modernized processes, sustainable infrastructure, thorough training and education platforms and programs, and regenerative fiscal structures.

What Are the Other Major Current Issues?

Food Deserts = no local or regional food production + no local or regional distribution.

In 2014, the contamination of tap water sources in central West Virginia sent already stressed food banks such as the Union Mission over the edge: the majority of the food they provide as relief in West Virginia comes from restaurants and other establishments in the Kanawha Valley.

Likewise, in 2016, the closing of the Walmart in McDowell County, WV severely limited food choice options for local residents, forcing them to purchase in bulk once a month, if they can get a ride to Tazewell County, VA, or Princeton, WV to the Walmarts there.

Typically, food deserts are discussed as low access to healthy food options by purchase. However, just as food production is a system, food consumption is also part of a system—and these two are linked. Addressing the issues of local food production diversity can be organized to also address issues associated with low healthy food access and consumption.

The recent COVID-19 pandemic revealed weak links in the U.S. food system already in play: the bulk of nutrition calories for most of the food in the United States is produced in only two states (California and Iowa) and comes to America's stores by only a few highways, and what is not supplied there, comes by ship from imports, reaching a recent all-time high of one-third of the food consumed in the United States.

During the pandemic, citizens who rely upon food relief or government benefits suffered most, and rural areas, where citizens struggle with food access even in "normal" times, were hit hardest.

The climate crisis exacerbates food supplies here in the United States as farmers in those Midwestern and Western States producing most of our domestic food struggle with new weather extremes. The slow pace of adaptation to these extremes in weather not only puts the food supply for the majority of Americans at risk, it also opens up new discussion and interest in places such as the Appalachians, which have an abundance of cheap land, significant rainfall, milder climate, and are central to U.S. centers of population.

PRODUCING FOOD AGAIN WHERE WE ARE: FOOD RESILIENCE, COMMUNITY RESILIENCE, CONNECTION TO HERITAGE, RESKILLING, SERVING LOCAL AND EXPORT MARKETS, ENDING OUR ADDICTIONS

As engaged citizens and practitioners in the Appalachian mountains, from our on-the-ground research and pilots, we have ascertained that additional natural resource-based agricultural sectors are ripe in our mountain environment for current investment for meeting current economic imperatives. These include: honey and bee product production, tree crops such as fruit, maple and other tree syrups, berries, and sustainable woodland harvest including medicinals and timbering. . .

Mountain Farming and Our Renewed Rural Identity

Our relationships to each other and to places evolve. Prior to the rise of coalfield identity in central Appalachia in the 1880s, people in subsections of this region identified in various ways: Southerners, pioneers, sharecroppers, with their families or kin, with their professions, with their states, and so forth. Prior to this, Europeans in the region were trappers, surveyors, or soldiers. Treading in and out of much central Appalachia place-based identity also is a strong masculine work identity: a coal mining family is defined as such because the men in that family work/ed in the mines. Every family member becomes part of the mining technology that is the coal miner himself.

Identities shift and change . . . and we can redefine relationships and regions to suit our current emerging modes and participants in work. With respect to the coalfields, many more people have left the coalfields than remain. Health care is the number one employer in most central Appalachian states. The emerging economic sectors highlighted in this white paper point to other potential family or work-based identities within grasp.

Regions ought to come together to form a common purpose and a common vision for this common ground through means such as a regional envisioning summit, branding and marketing campaigns, and also by being theorized by public intellectuals, and by lay citizen scientists and lay citizens.

Just as these prior "Appalachian" identities were not forever, other rural American identities are not forever. How else could or ought political, personal, economic, and cultural connections from rural regions or subregions be formulated?

The current transition from a coal region to a region having other meanings rests also in a cultural and identity conversion of "coal" families to something else as meaningful. Rural people can create an identity not tied to what was the biggest rural industrial intervention in their area. Models of land stewardship, linking to heritage and identity, bridging interests in rural life and hunting, fishing, and forest harvesting may offer ways to bridge the sympathies of local residents. Agriculture and the region's livestock production history offer a path to land connection that may resonate with remaining residents or attract in new ones.

In this new spate of identities also rests the potential for a more secure and prosperous region and an increase in the health of our citizens: agriculture can offer a meaningful path to movement, engagement with what constitutes healthy food to grow or sectors to steward, and new directions in community education that are wholistic, focused on work as meaning, means to economic, health, and land prosperity, and for family and community bonds.

Investment in Data, Research, Pilots, Infrastructure for Mountain Farming

To bring to viability the rich potential of mountain farming, we need the R-1 research universities and the federal and state agencies of the region to support:

- Agricultural, soil, production, erosion, fabrication, industrial engineering, and education research and data
- Modeling of regional sector upgrowth for infrastructure, education, population, logistics, water, and sustainability
- Investment for scaled sustainable mountain productions
- Investment in:
    - o High tunnel productions
    - o Vertical growing
    - o Urban indoor growing
    - o Repurposed indoor growing
    - o Aquaponics
    - o Repurposed industrial infrastructure for food production
    - o Repurposed surface mining sites for grazing, bamboo, poultry, berry production
    - o Agrihoods—communities with locally produced food at their centers
    - o Education from secondary to higher education
    - o Modeling of civic security and food security
    - o Models of increased land access, legal paths to land access, and land access for disenfranchised populations.

These investments could and ought to include USDA NRCS, private, and state partnering to scale:

- Farmer and cooperative productions.
- Farmers as cores for Agrihoods
- Housing and community planning
- Political buy-in
- Meat processing facilities
- Locally and regionally owned facilities run cooperatively, like that of Alleghany Meats in Monterey, VA

- Value-added food facilities
- Processing facilities to scale
- Research centers
- Research and data
- Regional conferences
- Regional Business councils
- Educational curricula and departments at community colleges, secondary schools, land grant universities
- Population decline mapping and analysis with economic sectors' potentials and staffing need
- Economic modeling of cottage industries versus sectors to scale
- Biological systems engineering testing
- Rural industrial site remediation
- Soil testing adapted to former brownfield sites repurposed for agricultural production
- Logistics mapping
- Youth development and retention. People remain the mountains' #1 export. A pipeline of young people in sustainable agriculture must be developed.
- Fresh water infrastructure
- Dissecting Appalachia into regions and what each region can excel in
- Small business creation modeling
- Job creation modeling.

Despite the need for research and modeling, solutions must be created with shorter-term, and, longer-term goals. Agricultural education and small business support must be able to take potential producers to viability quickly and efficiently without their being bogged down in dubious certifications or productions without clear and attainable waiting markets..Despite the need for research and modeling, solutions must be created with short-term and longer-term goals. Agricultural education and small business support must be able to take potential producers to viability quickly and efficiently without their being bogged down in dubious certifications or production without clear and attainable waiting markets.

Getting Ready for the Twenty-Second Century

We cannot stop at the work needed to catch up from the nineteenth-century to twenty-first-century imperatives for food sector build out and security, we must already turn our attention toward preparation for the twenty-second century.

Even in the best of predictions regarding sea level rise and rising heat in the U.S. South, the population of the United States is expected to radically shift in the next eighty years. In the Eastern Mountains of the United States, we must begin to prepare now for an influx of population to, and, a rise of food production in our region.

We envision a twenty-second century where the Appalachian mountains sustainably *teem with fruit and tree syrup production,* able to provide for local populations and for export to other states and regions.

We envision mountain farm production supporting *sustainable apiaries with honey and honey products to scale, providing far-reaching and region-supporting streams of revenue*. These mountains also can support bee tourism and provide sanctuary for pollinators, with pollinator corridors mapped and supported.

We envision small communities across the region reorganized into agri-hoods able to accommodate and to absorb population as people continue to relocate from the coasts and more heat-stricken sections of the United States.

We envision high tunnels, small farm infrastructure, and sites for sustainable woodland harvest dotting the landscape, providing work, food, income, and sustainable export and items for trade.

In contrast to relegating rural economies to large technical systems derived from heavy manipulation of natural assets such as the large-scaled nineteenth and twentieth-century interventions for economic wealth and community sustenance, for these twenty-second century oriented sectors to be sustainable, we must include working with local natural assets toward a regenerative and sustainable model of community health and fiscal solvency. Natural assets as well as people must be valued and centered. This effort must be intentional, planned, explicit, and with a space for many levels of participation and participants.

Conclusion

Research, Resources, Revealing, Redefining, Remaking

In *Big Rural* I engage with the development of a large technical intervention (as part of a large technical system) and with the construction of the Big Rural industrial space and its subsequent flattening of metis, democracy, and habitable environment. The large technical systems or other large processes that feed the metropoles largely reside in the rural industrial space, forming Big Rural. Rather than considered by the larger public or popular culture as on a continuum of modernity and industry with metropoles, the rural industrial space becomes construed culturally as the metropole's antithesis—a space both off-shored and ignored.

I have traced the role of scientists, engineers, and their allies and benefactors, intervening to create the rural industrial space. I have proposed ethical concerns for science and technology, and scientists and technologists, intervening in the rural as well for the corpora-science, government, and/or academic institutions functioning as their sponsors.

The Big Rural industrial space remains as subjected to categorization, classification, and scientific and technical intervention as the metropole space. The whitewashing of the rural industrial, and its large technical systems, obscures the ethnic and other human diversity of these spaces, erasing the rural space inhabitants altogether, unless a media subject as an exception or as an example of something in deficit through a narrative of poverty or environmental deficit or cultural or mental depravity. In the Big Rural industrial space, one is subjected to personal and global risk—the latter in the form of a sullied environment and the automation away of jobs, the former in having one's material and civic experience isolated and/or ignored or rendered as exceptional, rather than the rational result of industrial intervention.

This isolation of a person, and disciplinarily, of theoretical and cultural analysis, undermines the capacity for persons in the Big Rural industrial space to connect to other people in similar spaces and their plights. This disconnect keeps citizens in the Big Rural industrial space from more effectively organizing and responding to the major issues resulting from single-sector domination. One's individual liberty also becomes delimited by local single-sector dominance; one's democratic potential becomes sold away to corporate, Research and Development (R&D), and metropole imperatives. Often local government and the federal government fail workers and the environment in this rural space, either colluding in corruption or failing in providing adequate civic society support, and, failing to ensure policies first that support greater liberty.

To unveil the rural industrial I guided us through a case study, the Pocahontas Coalfield, to elucidate a model for reflection on the Big Rural as a concept, as a mode of rurality "too big to fail" in its totality and function. I move from assessing deficits and assets remaining in the rural industrial space to theorizing the role and results of the rural industrial space, to proposing actionable research and other agendas toward mediation and mitigation of the democratic and environmental deficits in that space.

The irony regarding the potential for reenvisioning and activating a different economy in the Pocahontas Coalfield, or other rural spaces subjected to single-sector economies, especially environmentally ravaging ones, remains that opportunities for new directions arise once large corporations and/or government entities have used up the resources once profitable, thereby gutting the stream of local government corruption.

In addition to this historical and theoretical conceptualization of Big Rural, I lay out how we, as citizens working together, may be able to break apart and remake "Big Rural" into another more metis-friendly and human, life-giving space.

By way of assisting others seeking to reconceive, to rethink, where they live, and how to begin to intervene toward change, I offer a summary of potential actions:

- Find and analyze the inciting intervention. How did the single-sector economy where you live come to be single sector? What stakeholders keep it in this state now?
- Examine the material interventions of the rural industrial site—not as an abstraction like "workers," "corporations," "science," "technology," "capital." What actual machines intervened or continue to intervene? What kinds of science and scientists? What kinds of engineering and engineers? Which research universities provide the research for this continued material intervention? What federal grants and agencies support this research? What corporations? What scientists and engineers work for the corporations involved in maintaining single-sector dominance? How could you intervene

with others in any way in this stream of science, technology, capital, and, could you propose other policies?

- Examine levels of liberty and technological intervention in your rural space. More plainly —does the depth of rural industrial intervention where you are match the depth of democracy deficit? And, if so, know that nothing historical or natural prescribes this. Know that you can work with others to find cracks or places to start to improve both material well-being in a range of kinds of wealth—personal, social, environmental, economic—and to increase freedom and democracy for local people even if voting still produces marginal change. Look again to the texts in the Bibliography and suggestions in the Endnotes in this volume for means to practice democracy beyond voting.
- Assess science and technology intervention realities with cultural or pop culture myths of this rural space. How are people in the media depicting your rural single-sector space? Why? Whom does this depiction benefit? What does it obstruct? What harm or good do these depictions cause? How could you alter or counter these depictions?
- Prescribe modes for change to increase democracy and to move away from losing work to automation, invisibility, necropower, and single-sector totalitarianism. Propose bold life-giving, environment-saving policy. Write your own manifestos, white papers, strategy documents, public proposals, and promote them.
- Describe modifications to the concept of Big Rural. Like other constructs in late modern capitalism, how does Big Rural manifest and morph from place to time to situation? What adaptations to response are needed to respond to Big Rural as it morphs?
- Describe and link changemakers already in motion in this site or adjacent site. Find the other people wanting to make a change, find common goals you can agree upon in some way (you don't have to see eye to eye on everything), and start.
- Engage in changemaking: strategy, policy, planning, pilots, rinse, repeat. Adjust. Fail and start again. Find your people to start with. Start where you are or move, for a while, to other places where you can apprentice to what they do to bring back to where you want to be.
- Engage in changemaking even, and especially, when and where the State fails. Find, create, and engage with grasstops—nonprofits, small municipalities, churches—and if lacking those, start your own affinity groups. Keep at it even when only a small group engage.
- Define group ethics and group relationship to democracy and the democratic citizen. Decide, definitively, that democracy and increasing the freedoms of the people with whom you engage are important, and, when considering a policy of any sort, how does that policy increase democracy

or freedom? Define how you practice democracy in your own group as you engage.
- Reflect, research, listen, engage, adjust, repeat.

One of the lessons while working in and on the Pocahontas Coalfield has been that in places with large funders, funding, infrastructure, strong community capitals or wealth, and overall line items available, that planning long-term makes solid sense. The surrounding support is there to plan and to work on models that seem like they could work "anywhere". In practice, however, in locations without these kinds of support, it can be best to first only commit to pilots as . changemakers must match up what is possible with what funding is available, assess, adjust, and then plan again.

More than one respondent I interviewed in, or who was associated with the Pocahontas Coalfield, spoke about working from the ashes of what has been left. With respect also to taking on new economic directions, when asked what was most needed in their town, one respondent answered with what is supposedly an old Chamber of Commerce joke:

> *The thing we need most to make change happen is about six good funerals.*

Rather than memoriam, let *Big Rural* as a text stand as an examination of how to consider more democratic intervention in highly technical and highly manipulated Big Rural, how to open the black box on Big Rural and its construction, and how to intervene in the rural with scientific, technical, and scholarly research and practices that support rather than reduce liberty and equity.

In the process of assembling this examination, I have engaged for hundreds of hours as a practitioner-researcher and social change agent in the region at hand. I have been fortunate to work with any number of groups in the Pocahontas and adjacent coal areas, who have "had the funeral" and are engaged in taking their economic and community visions in directions that could or do provide employment beyond coal; this is not to say that every town council member or community member is on board, but these groups hold steady in their path anyway, building and utilizing social capital as much as, if not more than, financial capital.[1]

Importantly, each of these entities, without having an explicit mission of increasing liberty for people who live in their communities, still engages with concepts of more choice, more work, more ways to make a living, more ways to connect to each other, and with potential economic drivers that do not depend on the kind of work that can be automated away.

Scalability stands at issue in each of their respective sectors, though here I want to make two points. Though tempting as it may be to insist that these small rural communities should not have to rely upon interstate and regional or even global commerce for their chosen economic direction/s, the reality on the ground in every case in the region of examination remains that nonlocal tourists, product export, and sales relations in metropoles largely determine local sector sustainability. Though food security remains also an issue in this region, and, being able to produce substantial amounts of one's own food in many cases may be possible as a family or in a community, complete self-sufficiency in any of these communities would be at most a kind of macho cultural fantasy. After all, despite the country and western anthem of a "Country Boy Can Survive," any number of material and technical items mentioned even in that song require trade with the non "backwoods."[2] My point being, largely, self-sufficiency of a location, a family, or a country has long been more fantasy than reality.

I assume that the people in the groups with which I engage have agency and the sophistication to understand that even their location connects to a global, or at least greater regional, marketplace. To this point, if any of the groups with which I deeply engaged receive the kind of financial backing, R&D support, academic research and energy, state and federal government push, educational commitment, and tax incentives offered the remaining and historic coal industry, then they too may be able to scale to meet the demands of the remaining and decreasing population for meaning, connection, community stewardship, and long term multigenerational work. *Big Rural* also stands as a call to action on these points.

Reviewing actual on-the-ground groups matters because though a single sector has come to dominate and to construct this rural industrial space, the single-sector's narrative of technology, work, social interaction, social capital, science, and research, does not remain the only work cum identity that has existed there or that can exist. While unlikely that a rural hinterland location will have many different and diverse economic sectors happening at once, it stands likely to have a range of businesses all dependent on local natural resources, but not the same in exact kind. For example, Williamson, West Virginia, known as the heart of the billion-dollar coalfields, also now has significant all-terrain vehicle (ATV) tourism for the Hatfield and McCoy Trail (depends on mountain trails), sponsors extreme sports events (mountain trails again), promotes local health and wellness activities, and has jumpstarted local markets and business related to agriculture (local land, some surface mined land). All of these in some ways are local resource dependent, but not all one sector.

Are three sectors diverse enough? If one believes, as I do, that a capitalist democracy structure such as ours privileges the democratic potential of a metropole citizen over that of a single-sector rural citizen, then, yes, three sectors are better than one. As I outlined in *chapter* 1, the totalitarian workplace cum single-sector community can weaken democratic practice. In Norway, strong democratic practice prior to their energy economy boom, along with their internal policies, shields Norwegians from the "resource curse" (though it does not shield spaces of Norwegian natural gas and mineral rights investment, like central West Virginia and Pennsylvania). Tthe means and modes of culture established in the Pocahontas Coalfield do not reinforce democracy as practice as demonstrated in my earlier discussions of corruption, government, corporate power, and democratic citizenship.

On-the-ground functioning models ought to be examined for additional areas for further economic, scientific, and technical research and collaboration. They ought to be examined to illustrate focuses on rural technology, scientific research, and economic advocacy. Said at a recent meeting of business owners in Bluefield, West Virginia—

> *any business in this context* [its myriad of social, economic, and other issues] *is social entrepreneurship.*

The main goal of my research into the single-sector economic sector of the Pocahontas Coalfield was to conceive of courses of action. In 2009, I had been convinced that technology of some sort or another would function as a savior to this kind of space. My understanding at that time of the role of technology in society was limited. Like many people I found myself enamored of digital technology, the internet, and what cell phones promised. At the time I analyzed the potential for intercommunications technology (ICT) such as cell phones and the internet to encourage democracy and to increase the potential for producers and people living at the bottom of the economic pyramid. I came to this research through a Schumpeterian lens of creative destruction—the concept favored in business research that old modes and industries fall away, and in their ashes, new, robust ones arise.

Let me spend a few minutes outlining how shifting from a technology and policy and innovation lens to continuing this research through a Science and Technology Studies of Society (STS) lens lifted the veil of techno-optimism from my eyes, and, cured me of a belief in historical determinism, that is, that there is a path to history that it marches along, with progress along each step, with the present always better than the past.

First, technology, history, industry, and for that matter, the economy, are constructed by humans and do not follow any natural laws that necessarily repeat or only follow particularly preset or predetermined patterns. Much space and scholarship has already been devoted to the debunking of historical determinism. More relevant here stands an examination of the practice begun from the historical and philosophical examination in the first chapters of this volume: if history does not repeat, if great and positive innovations do not arise necessarily from decline, if the economy is created by humans and not following an invisible hand, and if technology is not coming to save us, and also, not necessarily creating more equity or a better social fabric, then what ought we do? What ought we focus on?

Here is where I break considerably with what has become the expectation of modern scholars, especially in the social sciences and humanities. Although the field of STS grew out of the concerns of activist scientists with respect to the military industrial complex, these scientists, and later sociologists, forming alliances with historians and philosophers of science, rejected that science serve military interests.

Fast forward through an era where universities were once havens of scientific or scholarly pursuits for the sake of knowledge, at best, and places for educating the next generation of critical thinkers, at least.

In our current university-industrial complex, universities use STEM (science–technology–engineering–math) focuses to slash away fields and disciplines not tied to the pursuit of broader corporate interests and government funding imperatives (usually these two conflate, as corporations seek to externalize their R&D costs through government-sponsored corpora-science; they do not want to internalize the high cost of research). Science and the broader pursuit of knowledge for knowledge's sake cannot be commodified beyond that of tuition; knowledge must be tied to commodity and corporate and/or government contracts or grant-sanctioned production imperatives.

For someone wanting to pursue academic or scientific research not related to corporate or, energy, or defense research, universities no longer offer any reprieve. English Departments pitch their value in terms of producing more well-rounded and insightful and more communicative STEM practitioners. STS offers insight into scientific innovation and trains engineers to be more ethical. Each discipline must find its way of justification about STEM training, federal grant pursuit, or corporation donation or sponsorship. Where does one's discipline best serve as a handmaiden?

How do I know this? I can cite the work of scholars but I can also discuss this in terms of my experience for eight years of association with Virginia Tech both as student and employee. While departments in the humanities were being eliminated or shrunk, new multi-million-dollar institutes for "innovation" were being opened. I sat in on meetings while in hire abroad on behalf of Virginia Tech where foreign university presidents and Virginia Tech representatives openly discussed joint pursuit of federal U.S. grants. I worked for an on-campus center that worked closely with coal and natural gas companies. The following anecdote best illustrates university imperatives and attitudes.

Through my work in and on the Pocahontas Coalfield and region, I had come to this question: in a rural industrial environment, how might science assist in assessing how to rehabilitate soil for agricultural production that would be safe for human consumption? What protocols are needed? What matrix? Which science/s?

I trotted these questions before an array of scientists. Before I landed on someone who knew (Chris Barton at the University of Kentucky), I found myself sitting down with the Research and Development Department at Virginia Tech.

Despite being a National Science Foundation Graduate Fellow, which the gentlemen before me did not know, but would not have asked me about anyway, that is, they saw I worked in a center (at that time) and was not a head of any department. I explained the question at hand. They asked me if I had noted any federal RFPs (requests for proposals) on this topic. I had searched, and, no, there weren't any. One of them then declared: if there are no federal RFPs, then this must not be something worth knowing. I countered: they (the feds) may not know enough to know to issue an RFP. They thanked me for my time.

I will add to this—as an additional layer of cynicism, should scholars wish to pursue important knowledge that would be of benefit to a community regardless of its benefit to the government or corporate interests, the system of production of articles and books for an audience of only other academics that can be quantified to apply for tenure keeps these best and brightest minds occupied and detached. Articles and books for the lay public do not qualify for tenure. Thus, knowledge and theory by more novice scholars or thinkers in the academy stay obscure. Once tenured, the money for scholars to remain detached remains reasonable. The university frowns on activism, as does the larger academic community—especially in production, relegating theoretical work applicable or useful to lay public, "low" while that of interest to others in academe, "high." Serious academics produce "high" work.

The work of puncturing the ivory tower becomes about letting more and various kinds of people into this system as described above, rather than in flattening the tower. I offer two different kinds of experiences as counterpoints.

My reason for pursuing research on location in the region I studied, yet save for a handful of mainly tenured scholars, mostly in regional studies or sociology, I could have been anywhere. That is, given the depth of issues in the region, the university environment, with its hundreds of scientists and scholars, with its wealth and a huge cadre of students, might as well have been located on Mars, given how little time, thought, or resources were given to solving problems local or even adjacent to the Pocahontas Coalfield.

Who would stick it through to reshape the university environment now and its purpose? Ivory towers float above their locations, which serve as backdrops for recreational pursuits by students and maybe faculty. Students do not go to school where they are from. They leave the university community as soon as they are done with that school. Faculty concern themselves with the world of academe and tenure and obscure journals and insular conferences. The university chases corporate and federal money. Townies and locals and the region around continue to float on, diving and sinking in raging water around these wealthy and resource-full islands in the stream. Imagine if one research university turned some of its significant resources, even for a couple of years, to stilling the local waters, to preparing for or preventing the local floods?

Critics of asking academics or a university to devote larger portions of research and resources to addressing issues endemic and systemic in their locales may decry this imperative as thought patrol, politically correct, socialist, or any other kind of convenient slur. However, let's repeat a few lines from my experience at the Research and Development Office at Virginia Tech:

> *One of them then declared*: *if there are no federal RFPs, then this must not be something worth knowing.*
> *I countered*: *they* (*the feds*) *may not know enough to know to issue an RFP.*

In the United States, research imperatives already are being dictated in science and engineering by corporate partnerships and federal grant priorities (monopolies). Some scholarship and much administrative effort in the humanities are already devoted to proving the humanities' disciplines worth to the production of STEM students.

Colleagues, scholars, university administration, and the larger public can save the insincerity of shock at the concept of having a common mission or goal in scholarship. Right now, the research academy in the United States operates under the common goal of commodification of knowledge. This current U.S. model is not the only model possible.

Despite the majority influx of federal and state money, these goals of research universities are not put to the public to decide, except perhaps by proxy through boards no public has a hand in choosing. There is no democratic imperative in the United States for *locale* accountability or knowledge creation or sharing—no mechanism for real partnerships in terms of joint effort or fielding needs from the *locale* and its community. Agricultural Ed. departments and Economic Development may engage with localities, but they are subject to the tyranny of corporate and large federal grant imperatives like any other departments.

Now, some disciplines take a more active approach in the local—such as region-specific studies—Appalachian Studies being the one most pertinent to the case study region in this volume. At least in Appalachian Studies, while social change groups often have a place at the table with respect to conferences, and, some scholarly crossover exists into these groups in the region, most notable being in the issues of gas pipeline and surface mining protest, virtually no Appalachian Studies scholarship on or in the region crosses with technical scholarship.

Appalachian Studies scholars, or activist scholars, at best, black box technical concepts and the technical actors—the scholars do not partner or attempt to understand the technical processes, the technical stakeholders. "The economy" is blackboxed and its local actors are not revealed. Laborers are investigated, but the actual technical or material process of their labor largely remains unexamined in Appalachian Studies, especially as automation has reduced regional workforce in energy, and the work of natural resource extraction has become more and more technically complicated and reliant on microchips rather than pickaxes.

Let me venture one more critique that may get me blackballed in regional and Appalachian Studies, but which must be revealed. Just as other scholars in the humanities must volley and vie for tenure and rack up books or journal articles, largely unread by anyone outside their respective fields—a kind of worship of obscurity and reification of artistic or artisanal or labor union production embalms scholarship on Appalachia in an ambered past, a particular kind of "down to earth" "salt of the earth" White past.

This history scholarship in *Big Rural* also largely tells the story of a kind of White Appalachian past. Rather than the labor history story of capital (the coal barons, or what I termed "coal wranglers") and labor (the miners), a story also sanctified in Appalachian Studies, I expand the coalfield stakeholders backward in time to the lobbying for, and utilization of, science and scientists to make the case for the Pocahontas Coalfield intervention, and, I bring the case forward, and show the immense impact this science has had, and continues to have, and its range of current stakeholders: universities, DOD and DOE federal grants, conferences, energy companies, compliance scientists, geologists, engineers, mega-machines, the environment, energy policy . . . and I reiterate that like economists making the economy, Appalachian scholars also make Appalachia—and they leave the hardest hitting narrative, that of the massive science-based intervention of energy production in our region, as a culture, as a human endeavor, as a sociological tour de force, unexamined, unrevealed, dripping into a scholarly void, outside the amber.

Appalachian Studies largely ignores the impact of science and technology in the region as scientific practice, just as it ignores the scientists and scientific university departments that support the science necessary for these massive technological interventions. Scholars protest pipelines and interview protestors, but the scientists and science and engineering departments supporting these endeavors remain unexamined as Appalachian practices. Coal dominates the past in central Appalachia transition, but the current massive science and economic focus on natural gas as a cultural practice goes largely unexamined, other than some important scholarship on the influx of social issues surrounding the nearly all-male mobile gas industry workforce.

Of course, no discipline can do it all, but Appalachian culture is not ambered; science and scientific practice in the region both historically and currently merit massive examination, at least as much to equal their impact. Given the science backing these massive technological interventions currently impacting Appalachia, we need scholars also to take on including these interventions and their stakeholders in examinations of the makings of regional culture.

So, besides breaking science from corporate and grant-chasing control and tyranny, having universities focus majority research on solving local and locale issues, or at least devoting significant resources to that, and, my calling for regional and rural studies scholars and students also to investigate science and technology scholarship and support in the region as cultural practices, what else? What next?

I suggest a solid investment in the "low," as an invested and activist scholar. Although I am a scholar in STS, I also bring my scholarship, training, and practice to the table in literary analysis, creativity, and pedagogy.

I draw from but also beyond STS, Regional Studies, Rural Studies, Appalachian Studies, policy analysis, sociology, history of technology, and wander back into pedagogical possibility, literary analysis, and creative world-making.

I have long sought a book that would show where we have been, show new paths through the old system we inherited, even if we didn't create that path, and, that takes us on the ride of imagining other possibilities, even wild ones.

My own career and life constitute a ride of "high" and "low" culture;I see no "rural-urban" divide—it is all one society, one system. The false lines between high and low only serve as mechanisms for exclusion, with the "low" serving as a fount of ideas, change, and space for analysis by those gatekeeping the high. These false divisions maintain thesovereignty and order in the tower.

Time is precious. I want *Big Rural* to save you some time. Modify it how you wish. Take some lessons and leave others behind. I would like you to consider not only what you are against, but to deeply analyze what you think you are for. Go learn its history. Go find the actors that made that history happen and learn about them.

You are allowed to stand for something not only against something. Analyze (academics often stop here—the academy does not want you to advocate or activate). Advocate. Activate. Assess.

Big Rural is also my advocacy. YOU are as capable of this as anyone in Washington, and maybe more so. But you need policy based on research—but who gets to control the research? As I pointed out above—what questions get supported? Who pays for them to be supported? If not sanctioned by the state or the academy, then are those inquiries not worth pursuing?

Back to Lindytown, WV—how would we make it a place of life again? Just removing the murder weapons does not make it thrive, habitable.

My contention is if brownfields in Brooklyn and Detroit can be remediated for housing and urban gardens, then our environments in rural areas can also be remediated and renewed.

Jedidiah Hotchkiss' wager, its aftermath in this rural industrial area, and, the call for an engaged and socially responsive scholarship of space/place still stands. Hotchkiss dug in for thirty years to cross the finish line for a coalfield, but at the same time, he took a leap. A wild leap. But he took science with him.

And while some of the material in *Big Rural* will age, I want its spirit of inquiry, scholarship, criticism, and possibility to live on. I want you, the reader, whether academic, student, or layperson, to seize the permission to analyze, reflect, imagine, and create, as creating possibilities on the page, and with others in the polity, is as creative as any delimited creative or "artistic" pursuit.

What ought we take with us? How?

What will you take with you?

What next?

NOTES

1. Shannon Bell, "'There Ain't No Bond in Town Like There Used to Be': The Destruction of Social Capital in the West Virginia Coal Fields," *Sociological Forum* 24, no. 3, (September 2009): 631–57.

2. Mostly technology: shotgun, rifle, four-wheel drive, coal mines, .45, and one consumer good—Beech Nut, the chewing tobacco . . . and of course, Hank Williams Jr. mentions the "West Virginia coal mines," a highly technical rural industrial space. Hank Williams, Jr., "A Country Boy Can Survive" (January 1982).

Epilogue

In 2002 I stood in the "coal baron" created hamlet of Bramwell, WV amazed at the Victorian homes still intact amidst the nearby devastation of depopulated and abandoned Route 52. The cobblestones and stately houses reminded me of what I had seen in sections of California, or, in small college towns in New England. Bramwell stood in stark contrast to the disinvestment and deindustrialization of most of the rest of the Pocahontas Coalfield.

In 2006, hot on the heels of all the "new creative economy" buzz surrounding the potential for artists and creatives to turn around deindustrialized places like Detroit or Pittsburgh, I went so far as to explore what might be possible if I attempted to start an art school in Mercer and/or McDowell County in West Virginia. As I explain in the Preface to *Big Rural*, in 2009 I picked this up as a research journey that led me not only to writing the volume you are reading but also to co-found a nonprofit that supports training and growth in local potential scalable sectors.

Recent opportunities and life events have pushed me to reconsider the option of a community art space in McDowell County as a democratic endeavor. First, my father, who hailed from McDowell County, passed, leaving behind a tremendous archive of creative output in graphic design, board and other game design, creative nonfiction, and painting. An "outsider" artist in that he did not have formal training and did not subject his creative output to the feedback and critical loops of a larger creative community, his work still expresses a particular sensibility of social and class critique worth exploring. My childhood also was punctuated by his continuous flow of stories of his work in McDowell and other West Virginia counties when he worked as an insurance investigator in the 1960s, and, from his childhood and young adulthood there.

The summer before he passed, I had talked to him about the opportunity to buy The Big Four Motel in the Town of Kimball, WV—a development project pointed out to me and others while touring with a real estate developer considering what could be worth developing in McDowell County. Usually underwhelmed by such a proposal, my father responded with enthusiasm as he knew the whole history of this motel, why it was named what it was named, and could see its potential.

A couple of months later on his deathbed, he asked me about the motel and encouraged me to pursue it. Two months after he passed, I closed on it. Since then, both the practical and greater purpose of the space have become clearer to me.

Recently, academics have engaged in a robust discussion of the extraction of knowledge from people in a peripheral space. McDowell County is a regional favorite journey by journalists, filmmakers, TV personalities to extract story, as any sojourn there often comes preloaded with a manufactured "authenticity" visually and in media, as the tropes, images, archetypes, already run deep in U.S. culture.[1] In addition to the Big Four Motel as a motel and an engine for local reinvestment, what if I also opened a creative space there, and/or opened the space to creative production? What if the motel also offered space for folks from McDowell County to engage creatively with themselves, their neighbors, and with folks from other places?

The idea felt difficult, and in some ways, dangerous: How could I work with other people to make space for challenging art from McDowell County residents and in conversation with them? With so much other need there, was this idea frivolous?

Since *Big Rural* is also about metis and the democratic citizen, piercing the flattening of metis is not only about reintroducing local foodways production, but also about reintroducing avenues for local cultural and thought possibilities.

How do places talk to themselves? Where?

If McDowell County could talk to itself honestly, what would it say? If people were not cut out of the conversation, what would they say? What intimate knowledge would they reclaim? Repossess? What new knowledge would they invent, invest in?

What would McDowell say to people in places in similar straits?

What would that engagement look like?

Through the vehicle of his white-collar work in the 1960s and from growing up in and around McDowell County, my father had developed an intimate knowledge—in and out of its hollers, interacting with all of its people. As a child and teenager with nowhere to take his artistic talents, his became an interior mind—analytical and artistic. But, with nowhere to go, it became entirely an expression of an interior monologue—not a dialogue with the viewer, the reader, the spectator.

Metis is not only of that interior mind, but also of shared and negotiated intimate knowledge.

McDowell is not alone.

McDowell is not in a fishbowl and has every much as a right as any metropole space to speak to itself and with others.

I cannot predict what art or creative output people want or need from The Big Four Motel, or which will resonate with the community there and put it into dialogue with people from elsewhere.

Reclaiming oneself or one's community away from being a "subject" under the control of a king-like entity requires more than reclaiming economic control and foodways control, but also connection and engagement in cultural, thought, and civic discourse toward re-becoming or becoming a citizen.

As throughout the rural space with Big Rural intention and intervention everywhere, no stone left unexamined, no land left unsurveyed for actual or potential exploitation, what could a creative space wedge open and hold the door for? What state of mind or creation or challenge in perception brings a person beyond the state of exception?

Any of this is tall order for an art/creative space and concept still under development, but reacting not only against the environmental and economic but also the cultural extraction of a rural industrial place like the Pocahontas Coalfield strikes me as two sides of the same coin to remake the *Big Rural* space as a place that is life-giving and sustaining.

The Big Four Motel and engagement around *Big Rural*—for now, for me, that is also what's next.

NOTE

1. A recently culled set of samples from the internet:

"Appalachian Christmas" by *Saturday Night Live*: https://www.youtube.com/watch?v=lNZe13qDNpo&fbclid=IwAR3OqWbOz2s-_L1OZW1jhhda9EGL3GleydKC0He6hZTk4vATaHf4SboEsBM

Images by Stacey Kranitz: https://www.google.com/search?q=stacy+kranitz&rlz=1CAXLEN_enUS925&sxsrf=ALeKk01n7q808T_Dre4xgn1fY2vSYpEe5g:1606938989581&source=lnms&tbm=isch&sa=X&ved=2ahUKEwiF8JidirDtAhVlSTABHWuzDOoQ_AUoAnoECBQQBA&cshid=1606939186926387&biw=1300&bih=580

Images by: Bruce Gilden of *Vice*:

https://www.vice.com/en/article/vdxvx8/two-days-in-appalachia-0000687-v22n7?fbclid=IwAR1wwaBjDlkfm3EudG0ntFwADWy4Docq2nY30KLneCD8nshQZW6m0O04S5A

Washington Post (many articles about McDowell County): https://www.washingtonpost.com/local/education/teaching-election-west-virginia/2020/11/29/b8e40a38-2b3a-11eb-92b7-6ef17b3fe3b4_story.html

New York Times:

https://www.nytimes.com/2014/04/21/us/50-years-into-the-war-on-poverty-hardship-hits-back.html

The Dancing Outlaw:

https://www.reddit.com/r/Music/comments/1dck2v/the_dancing_outlaw_on_an_episode_of_roseanne/

Anthony Bourdain:

https://explorepartsunknown.com/destination/west-virginia/

Mike Rowe:

https://www.herald-dispatch.com/_recent_news/season-two-finale-of-mike-rowes-facebook-watch-show-spotlights-southern-wv/article_655ffcee-aaf1-11e8-a2fb-5fbf8013007f.html

Christian do-gooders:

https://www.christianapp.org/blog-news/christian-appalachian-project-recognizes-feed-children-champion-appalachia

"Buckwild" TV Show:

https://www.mtv.com/shows/6pmm6h/buckwild

Ron Howard and Hillbilly Elegy:

https://www.newyorker.com/culture/the-front-row/the-silent-political-messaging-in-ron-howards-hillbilly-elegy-adaptation?source=search_google_dsa_paid&gclid=Cj0KCQiAk53-BRD0ARIsAJuNhpvXc2ssscg7DSu_wK2lkprbjIwEVADER4vCPQUe7POQqVn5sP3RNOgaAuQ9EALw_wcB

American Hollow:

https://lists.h-net.org/cgi-bin/logbrowse.pl?trx=vx&list=h-appalachia&month=9912&week=a&msg=EhZiNugX4DbSg%2BeEf2jqLg&user&pw&fbclid=IwAR2k-VIy3t7h_T6y_4Jg7faoB4e_9nJdKQUix8zl5334577Ja1VcOMj7NZc

https://www.washingtonpost.com/archive/lifestyle/1999/11/29/hollow-shallow-too/ba2710b5-fbdb-4153-99e5-c30827ffac75/?fbclid=IwAR0ZTlRyqqDBQpk_63aSNyKUhQLK1m-2qe1cQMOB3WljFahOZlGffOCXIGQ

Appendices

APPENDIX A

Selected excerpts from landing pages for Mining Engineering programs in the central Appalachian region demonstrate the imperatives for research, career development, and contacts for corporate jobs.

West Virginia University's Mining Engineering's webpage outlines key faculty research.[1]

> Faculty members have wide-ranging expertise and are active in research in specialties . . . within mining engineering, including mine systems, rock mechanics and ground control, mineral/coal processing, and mine health and safety. Faculty members consult with industry and governments around the world and are committed to mentoring students who are interested in pursuing careers in mineral sciences.

University of Kentucky's Department of Mining Engineering's landing page promotes earnings and career potential.

Virginia Tech's Mining Engineering pages, as of 2016, stressed its connections to corporate partners with CONSOL Energy featuring prominently.

The heads of the department are shown with Virginia State politicians.

The former head of the Virginia Center for Coal and Energy Research, as of 2022, sounds like he is mimicking my analysis of Southwest Virginia from 2016[2]:

> But as he steps away from the university, Karmis' attention has turned from mitigating issues related to mining operations in Appalachia to helping the region recover financially from 30 years of declining coal production.

In 1990, coal employed more than 10,000 Virginians, according to the state Department of Energy. By 2020, fewer than 2,000 of those jobs remained. . . . In some parts of Southwest Virginia, more than 30 percent of children live in poverty, according to the Appalachian Regional Commission.

"Southwest Virginia is in a difficult situation," Karmis said . . . "It's not easy to replace jobs lost that paid $80,000 a year."

And the region still has a lot to offer employers, Quillen said. The perception of coal miners as guys working with picks and shovels is woefully outdated. . . .

"It's very sophisticated, mechanized, and there's so much computerized electronics," Quillen said. "The talent is here, from electricians to mechanics to welders, and they have a tremendous skill set. We just have to figure out how to attract alternative jobs to the southwest region."

Karmis said he sees some hope on the horizon, including a recent federal working group report that identified $38 billion in government funding that could be diverted to revitalization projects in distressed former energy communities, primarily in Appalachia.

And he said interest continues to grow for commercial projects in the region, including energy storage, harvesting rare earth and critical minerals from coal waste, mine reclamation, and renewable energy, such as solar and hydrogen production. . . .

And, last year, a United Nations-affiliated group recognized a department course in sustainable mining taught by Associate Professor Emily Sarver and a handful of collaborators in the United States, Colombia, and Chile, as a model of sustainability-focused education.

APPENDIX B

To round out this toe in the water of technical spokes in and out of the Pocahontas Coalfield, the associated occupations of coal mining and recent real hired numbers—here is a buffet of jobs far beyond that of miner. Refer to the Bureau of Labor Statistics for the full list.

SOC Major Groups in NAICS 212100—Coal Mining[3]:

- 00-0000 All Occupations
- 11-0000 Management Occupations
- 13-0000 Business and Financial Operations Occupations
- 15-0000 Computer and Mathematical Occupations
- 17-0000 Architecture and Engineering Occupations
- 19-0000 Life, Physical, and Social Science Occupations
- 29-0000 Healthcare Practitioners and Technical Occupations
- 33-0000 Protective Service Occupations
- 37-0000 Building and Grounds Cleaning and Maintenance Occupations
- 41-0000 Sales and Related Occupations

- 43-0000 Office and Administrative Support Occupations
- 47-0000 Construction and Extraction Occupations
- 49-0000 Installation, Maintenance, and Repair Occupations
- 51-0000 Production Occupations
- 53-0000 Transportation and Material Moving Occupations

APPENDIX C

Approaches for Community Based Technology and Innovation in Economically Challenged Communities.

Full List of Questions:

1. How would you characterize the state of your community currently?
2. Briefly describe the recent history of the community where you reside and what factors contributed to its current state.
3. What activities aimed at community revitalization do you currently participate in? Please describe the activities and your role.
4. What are the goals of these activities? Please describe any progress or challenges toward meeting these goals.
5. Describe a proposed activity for community revitalization that you thought might be far-reaching or unusual. Assess its viability in your community for success. What brings you to this conclusion?
6. Describe the most innovative or creative community revitalization activity, project, or action you have encountered in your community. What makes it innovative or creative?
7. What is your community's most pressing struggle?
8. What are your community's strongest assets? Why?
9. How would you characterize the people that actively participate in community revitalization activities?
10. What kinds of people would you like to see involved in the community revitalization activities that are not?
11. What material, infrastructural, technical, educational, artistic, or civic resources do you think your community currently has?
12. What material, infrastructural, technical, educational, artistic, or civic resources do you think your community needs that it doesn't currently have and why?
13. If you could envision anything for this community, what would you like to see occur? What would you like to see it have?
14. What makes this community a community?
15. Whom else should I interview about community revitalization or resilience in your community?

APPENDIX D

The full data set can be found on the companion website to *Big Rural*. This data set includes counties in Virginia, West Virginia, North Carolina, Tennessee, and Kentucky, beginning with Montgomery County, VA, and ending at McCreary County, KY.

APPENDIX E

USAID is the lead U.S. Government agency that works to end extreme global poverty and enable resilient, democratic societies to realize their potential:

> The Conquest of Poverty
> Assistance to Foreign Countries
>
> U.S. foreign assistance has always had the twofold purpose of furthering America's interests [What in America's interests is at stake? Is coal or full democracy for Americans what is at stake? Is this cost benefit analyses democracy, where metropolitan inhabitants get more democracy if someone in Lindytown, WV gets less?] while improving lives in the developing world. USAID carries out U.S. foreign policy by promoting broad-scale human progress [Large-scale?] at the same time it expands stable, free societies, creates markets and trade partners for the United States, and fosters good will abroad.
>
> Spending less than 1 percent of the total federal budget, USAID works in over 100 countries to:
>
> - Promote broadly shared economic prosperity;
> - Strengthen democracy and good governance;
> - Protect human rights;
> - Improve global health;
> - Advance food security and agriculture;
> - Improve environmental sustainability;
> - Further education;
> - Help societies prevent and recover from conflicts; and
> - Provide humanitarian assistance in the wake of natural and man-made disasters.[4]

NOTES

1. Wvustatler. "Mining Engineering: Research." Mining Engineering | West Virginia University. Accessed August 20, 2023. http://mine.statler.wvu.edu/research.

2. "Former Energy Center Director Prospects for Appalachian Opportunity." Virginia Tech News | Virginia Tech, February 3, 2022. https://news.vt.edu/articles/2022/01/former-energy-center-director-turns-attention-to-revitalizing-ap.html.
3. "Coal Mining—May 2022 OEWS Industry-Specific Occupational Employment and Wage Estimates." U.S. Bureau of Labor Statistics, April 25, 2023. https://www.bls.gov/oes/current/naics4_212100.htm.
4. "Organization: About Us." U.S. Agency for International Development, January 17, 2023. https://www.usaid.gov/about-us/organization.

Glossary

Reviewing this glossary may assist with unfamiliar terms or words with less familiar theoretical uses.

A

agrihoods: a housing development or similar location where a farm, farmer, or group of farmers (see the definition for "farmer" below) is also located and produces significant amounts of food for that development, neighborhood, or town. The model varies from a farm turned into a housing development but a portion of the farm still remains in production for that development to a farmer or groups of farmers growing food to address a dearth of hyper-localized food production. In the case of introducing food production into a community without it currently, an agrihood could be developed on a vacant lot, at the edge of cities, or by urban farmers utilizing fallow lots.

B

black box (n). Black boxing/black-boxed: to use a concept or a term, especially in academic writing, that is taken for granted and not defined precisely, or, examined for all that terms really indicates. In terms of this volume, that could be a term like the "economy" or "capital." Often academics write about these two concepts without revealing the people behind the economy and capital (investment institutions, and banks and federal agencies, etc.) that make up the economy or capital, or, at least make significant decisions about them that affect the rest of us.

brownfield: land that has been contaminated through industrial use and needs to be rehabilitated or remediated in order to be put to nonindustrial uses safely.

C

capital (as animated) (Introduction, p. 19): all the people and institutions that can lend, give, or invest money.

complicate: reveal all the ins and outs of a concept, idea, plan, problem, and so on.

- "Complicate rural as a construct" (Introduction, p. 25)

constructs (n): a term used in academic writing which ought to be defined precisely and explicitly.

- technology constructs

constructs (v): to create, to establish

contend (v): to assert

- "Contend theoretically"

cultural examination: looking at an issue from the way people knowingly or unknowingly create this issue, problem, or artifact specific to their culture.

cultural hegemony: when one group's culture dominates and is used as the standard by which other ways and means of social action or behavior are judged, valued, or acted upon.

culturally manufactured: that people create the society they live in, including the professions and industries in a society.

D

deadname: especially when someone that identifies as nonbinary or transexual has chosen a name that better suits their gender identity and someone else accidentally or on purpose continues to call that person by the name they no longer use/identify with. To deadname often is used to diminish the will of the person to identify with a name central to their identity rather than a name given to them by someone else.

demarcate: in this case, all the social, cultural, and political things we humans create and have created to show we are citizens of a state and not just subjects. For example, back in the Middle Ages in Europe, most people were subject to the laws of the monarchy and did not get a vote in any of these laws. Becoming a "citizen" has not been linear in Western culture, and people in the United States (the country most examined in this volume) continue to negotiate through laws, courts, practice, how much control any of us has over our lives. Voting is often "demarcated" as the pinnacle act of acting like a citizen; however, this is a very limited action compared to all the ways that power in our society can be affected through other means such as money, lobbying, back room deals, strikes, the military, the police, and so on.

- Demarcate "citizens"

disciplinarity: the last 200 years have seen an explosion of knowledge being divided into fields, disciplines, subfields, and specialties. What we take for granted now as a range of disciplines from biology to physics to chemistry to psychology, prior to the creation of many scientific instruments that could measure our natural world and us as humans, were much more theoretical and fell together under broad terms like "philosophy."

dissertation (chapter 2, p. 3, para. 3): in academia, usually a written product by a student studying for their doctoral degree, in which case the student makes a theoretical examination of a topic, and a panel of academic advisors decides whether this product meets the standards for being a doctorate of philosophy in that subject.

E

environmental cost shaftings: environmental costs are those that it would take to make a place like a brownfield habitable again or that deal with contamination, waste, trash, pollution from an industry. "Shafting" = colloquial—to shaft is to fool or to steal something in a way that the person being stolen from may not realize it or is powerless to stop it.

expository (interlude 1, p. 1): having to do with writing out an explanation.

F

farmer (n): a person that grows food, usually for sale, and for consumption by other people. The focus in this volume is on farmers that engage in production agriculture, and farming can indicate a range of food producing interventions from field crops to maple syrup production to honey production to growing indoors.

field disciplinarity: the creation of different fields and where one ends and the other begins, such as, biology means all the science that does these kinds of things, whereas chemistry is a science that means another set of things, and so forth.

foreground (v): to give knowledge or facts to help understand a topic.

G

grasstop: a person that often runs a grassroots organization or is in a midlevel position of power in a municipality, often in charge of social or business programming. In the case of this volume, I interviewed grasstops that included people that started arts organizations, heads of Chambers of Commerce, nonprofit executive directors, farmers, city managers, county commissioners, and so on. Whereas grassroots often means a movement that comes from regular citizens not getting paid to do social change; grasstops would be people elected or getting paid to do social change or manage social interactions, and, who, often may have more of an overview of a social, economic, or political situation than those deeply involved in creating change at the same level as non-grasstop citizens.

H

high tunnels: similar to a greenhouse or hoop house—a covered structure used most often for growing plants.

holler (n): a small valley between two hills or mountains. Holler is the colloquial version of "hollow," but the word means much more than that, and, to mountain people in the U.S. South, indicates a kind of social organization or neighborhood.

I

interests (in the Marxian sense) (Introduction, p. 26): roughly, all the trappings, social, political, economic, that influence or belong to a particular social class and which define what may be optimal outcomes for that social class. This is a large topic in social and economic theory, but this definition will get you through what is intended in this volume.

intervention: creating something in society.

socio-technical intervention (chapter 2, p. 1): a creation in society that also utilizes technology—think, our energy grid or our highway system.

highly technologically intervened (chapter 2, p. 4): engineering or technical constructions or creations.

large technical intervention: again, a creation in society that also utilizes technology, such as our energy grid or our highway system.

L

living spaces: in this case not just our homes or apartment buildings, but our towns, villages, neighborhoods, "hollers."

M

materiality: the totality of all the tangible, touchable objects and things in life and the places we use that define our culture.

material culture: all the tangible, touchable objects and things in life and the places we use that define our culture.

material space: places we can see, touch, live in, inhabit, frequent that define our culture. In the case of this book, a material space could be a deep mine, a surface mine, a town, a holler, and so forth.

metis: local, specific, intimate knowledge. In the case of this volume, especially of a place.

O

obtain (Example: chapter 2, p. 19 "These relationships obtain." chapter 2, p. 25 "same economic geography obtained."): remain in play, or, as is usual.

object boundaries (boundary objects): objects that can mean different things depending on their context or who and where they are used; in science, they can also be all the objects that help define and make one kind of science distinct from another kind of science.

offshored internally: to offshore usually means to move some kind of factory or other production to a place with lower wages for workers or no taxes on business income. Usually, this is done, in the U.S. context, by moving this factory or investment to a place not subject to U.S. laws. However, if somewhere else in the United States also has lower wages or fewer taxes on business income, it seems reasonable to assume that some internal offshoring also occurs in the United States.

P

panopticon: a building created so that its contents can be viewed from a single location. Think of the viewing tower of a prison. There is a rich discussion in philosophy about the actual physical presence of a panopticon, and also about the kinds of things in our society that function to make us "behave" (like the threat of being arrested for breaking a law) even if we are not being observed all the time like in an actual prison.

regional panopticon: controlling a region from one place as if that place controlled the whole region.

people as technologies (chapter 2, p. 31): maybe most simply put, if a machine can replace people doing some kind of work, then before that machine is invented, the people themselves were functioning as a kind of technology.

polity: organized society, including the institutions that govern it.

postulation: to believe or assume to be true.

practice: acts in society, aka, what you do in society, and the acts in each of these following uses of that term:

Academic practices
Negotiated practices
Research practice
Social change practice

protocols: "scientific and technical protocols": the usual acts and orders of a scientific or technical discipline.

R

raw material production areas: places where natural resources are grown or extracted for use.

RFPs: requests for proposals—what the U.S. federal agencies use to indicate they want to fund some kind of research or project. Then, scientists, engineers, academics, and nonprofits put together proposals in response to these RFPs.

rural industrial space: a place where natural resources are grown or extracted for use.

rurality: all the social, cultural, technical, environmental facets specific to a specific rural place. The concept of rurality asserts that you can't paint all rural places with the same brush; there are many differences among them, and, if you are going to try to intervene in some way, then you need to take very local variations into account.

S

saliency: relevance to a subject.

social constructivism: that learning is reciprocal and that we learn through collaboration, and, that is also how we create what we create in terms of our culture.

soil path agriculture (chapter 5, p. 14): focusing on the primacy of soil to support life on our planet.

space: not just a place, but all the things that go along with making that place particular—the environment, climate, weather, people, culture, language, kinds of work, kinds of play, etc.

surface mining adjacent living spaces: towns, villages, neighborhoods, "hollers" near surface mined sites, usually predating the surface mining occurring.

systems: in the usage here, things in our society that can affect large groups of people and are organized to do just that: think education system, transportation system, tax system, electrical grid, logistics for shipping and receiving, the distribution of food to supermarkets, and so forth.

- Large technical systems: big systems that can function so well we forget they are systems, especially systems like our energy grid, or, large-scale agriculture.

T

Taylorism: one of the first attempts to make work life and its organization more rational and streamlined.

teleological historicism: the false concept that things are getting better with time or that history demonstrates humans are on a path of progress.

text: a book, volume, theory, journal article, maybe even one worth studying or thinking about.

theorizing: to come up with why you think something is the way it is.

tropes: figures of speech, but which often are not literal.

U

Ur: the earliest known of a thing, kind, place, people, and so on.

- An "Ur" (proto, primal, or original) Claim" (chapter 4, p. 35)
- "Ur-Appalachian Claim" (chapter 3, p. 11)
- Ur-narrative
- "Ur-right" (chapter 3, p. 12)

V

value added food facilities: where food from a field or livestock is changed into products that often command a higher price and have a longer shelf life than whole food ingredients might. For example, while hamburger meat is somewhat processed, thus it goes up in value per pound from just buying a cow, beef jerky is highly processed and has a long shelf life, and, is sold for much more per pound than hamburger. Beef jerky is value-added beef.

W

wooly back: people outside Liverpool, referring also to those who participated in the dockers' strike.

Bibliography

Abrams, Lynn. *Oral History Theory, Second Edition.* Routledge, 2016.

Adams, Sean Patrick. *Old Dominion, Industrial Commonwealth: Coal, Politics, and Economy in Antebellum America.* Baltimore: Johns Hopkins University Press, 2004.

Ahlstrom, David. "Innovation and Growth: How Business Contributes to Society." *Academy of Management Perspectives*, 24, no. 3 (2010): 11–24.

Akenson, Donald. *God's Peoples: Covenant and Land in South Africa, Israel, and Ulster.* Ithaca, NY: Cornell University Press, 1992.

Alexander, Peter, Aiden Berri, Dominic Moran, David Reay, and Mark D.A. Rounsevell.

Anderson, Benedict. *Imagined Communities: Reflections on the Origin and Spread of Nationalism.* Verso Books, 2006.

Atkin, Emily. "Al Gore's Carbon Footprint Doesn't Matter." *The New Republic.* August 7, 2017. Accessed on July 13, 2–23. https://newrepublic.com/article/144199/al-gores-carbon-footprint-doesnt-matter

Auty, Richard. "Natural Resources, Capital Accumulation and the Resource Curse." *Ecological Economics* 61, no. 4 (March 15, 2007): 627–34.

Barry, Dan. "As the Mountaintops Falls, Coal Town Vanishes." *New York Times*, April 12, 2011. https://www.nytimes.com/2011/04/13/us/13lindytown.html

Barton, Christopher D., Interview by Crystal Allene Cook Marshall. July 2016.

Beck, Ulrich. *Risk Society: Towards a New Modernity.* London: Sage Publications, 1992.

Bell, Shannon. "'There Ain't No Bond in Town Like There Used to Be': The Destruction of Social Capital in the West Virginia Coal Fields." *Sociological Forum* 24, no. 3 (September 2009): 631–57.

Ben Hassen, Tarek, and Hamid El Bilali. "Impacts of the Russia-Ukraine War on Global Food Security: Towards More Sustainable and Resilient Food Systems?" *Foods* 11, no. 15 (2022): 2301. https://doi.org/10.3390/foods11152301

Berkes, Howard. “Advanced Black Lung Cases Surge in Appalachia.” *All Things Considered*, December 15, 2016, http://www.npr.org/2016/12/15/505577680/advanced-black-lung-cases-surge-in-appalachia

Bijker, Wiebe, Trevor Pinch, and Thomas Hughes. *The Social Construction of Technological Systems: New Directions in the Sociology and History of Technology*. Cambridge, MA: M.I.T. Press, 1987.

Bittman, Mark. “Everyone eats there.” *New York Times Magazine*, October 14, 2012, https://www.nytimes.com/2012/10/14/magazine/californias-central-valley-land-of-a-billion-vegetables.html

Bloor, David, Jan Wolenski, and Matti Sintonen. “Sociology of Scientific Knowledge.” In *Handbook of Epistemology*, edited by Ilkka Niiniluoto, 919–62. Boston: Kluwer Academic Publishers, 2004.

Bonano, Alessandro, and Douglas Constance. *Stories of Globalization: Transnational Corporations, Resistance, and the State*. Penn State University Press, 2008.

Bossidy, John Collins. Toast at the 1910 Holy Cross College Alumni Midwinter Dinner.

Bowker, Geoffrey C., and Susan Leigh Star. *Sorting Things Out: Classification and Its Consequences*. Cambridge, MA: MIT Press, 1999.

Breslau, Daniel. “Economics Invents the Economy: Mathematics, Statistics, and Models in the Work of Irving Fisher and Wesley Mitchell.” *Theory and Society* 32, no. 3 (2003): 379–411.

Brown, David L. and Louis E. Swanson. *Challenges for Rural America in the Twenty-First Century*. Penn State University Press, 2010.

Brown, Jeffrey R., and Jiekun Huang, “All the President’s Friends: Political Access and Firm

Brown, Mark B. “Politicizing Science: Conceptions of Politics in Science and Technology Studies.” *Social Studies of Science* 45, no. 1 (December 23, 2014): 3–30. https://doi.org/10.1177/0306312714556694

Browne, William P. *The Failure of National Rural Policy*. Washington, DC: Georgetown University Press, 2001.

Bush, Vannevar. “Science The Endless Frontier.” Director of the Office of Scientific Research and Development, July 1945. https://www.nsf.gov/od/lpa/nsf50/vbush1945.htm

Butler, Judith. *Bodies That Matter: On the Discursive Limits of “Sex.”* New York: Routledge, 1993.

Callon, Michel, ed. *The Law of the Markets*. Hoboken. NJ: Wiley–Blackwell, 1998.

Campbell, Hugh, Michael Mayerfeld Bell, and Margaret Finney, eds. *Country Boys: Masculinity and Rural Life*, Penn State University Press, 2006.

Carr, Patrick, and Maria Kefalas. *Hollowing Out the Middle: The Rural Brain Drain and What It Means for America*. Boston: Beacon Press, 2010.

Caswell, Margriet, Keith Fuglie, Cassandra Ingram, and Catherine Kascak, *Adoption of Agricultural Production Practices: Lessons Learned from the US Department of Agriculture Area Studies Project*. Agricultural Economic Report No. AER792. Economic Research Service, U.S. Department of Agriculture, January 2001, https://www.ers.usda.gov/publications/pub-details/?pubid=41202

Caudill, Harry. *Night Comes to the Cumberlands: A Biography of a Depressed Area*. Boston: Little, Brown and Company, 1963.

Center for Media and Democracy, "Massey Energy," *SourceWatch*. n.d., https://www.sourcewatch.org/index.php/Massey_Energy

Clark, Amy D., and Nancy M. Hayward. *Talking Appalachian: Voice, Identity, and Community*. Lexington: University Press of Kentucky, 2014.

"Coal Mining—May 2022 OEWS Industry-Specific Occupational Employment and Wage Estimates." U.S. Bureau of Labor Statistics, April 25, 2023. https://www.bls.gov/oes/current/naics4_212100.htm.

Congressional Research Service, "Department of Defense Energy Initiatives: Background and Issues for Congress," by Schwartz, Moshe, Katherine Blakeley, and Ronald O'Rourke. June 26, 2012.

Contrera, Jessica. "In a Place of Need, an Unhealthy Contradiction: They Are Poor, Sick and Voted for Trump. What Will Happen to Them without Obamacare?" *The Washington Post*. March 11, 2017.

Cook, Crystal. "Replicas," southernhum.com, 2006.

Cook, Nicholas, Leigh-Ann Krometis. and Emily Sarver. "Inventory of Bacterial Impairments in Central Appalachia." In *Environmental Considerations in Energy Production*, edited by John R. Craynon, 214–27. Englewood: Society for Mining Metallurgy. and Exploration Inc., 2013.

Cook, Rob, "Rankings of states that produced the most food." *Beef2Live*. Updated July 24, 2023. https://beef2live.com/story-states-produce-food-value-0-107252

Coulter, Dana. "The Fall of the Maya: 'They Did It to Themselves.'" *NASA Science*. October 6, 2009. https://science.nasa.gov/science-news/science-at-nasa/2009/06oct_maya

Cowie, Joseph and Joseph Heathcott, *The Meanings of Deindustrialization, in Beyond the Ruins: The Meanings of Deindustrialization*. Edited by Joseph Cowie and Joseph Heathcott, Ithaca: ILR Press, 2003.

Cunningham, Rodger. *Apples on the Flood: Minority Discourse and Appalachia*. University of Tennessee Press, 1991.

Demmitt, Jacob, "Virginia Tech Planning 300-Acre 'Automation Park' to Test Drones, Self-Driving Cars." *The Roanoke Times*. August 29, 2016. http://www.roanoke.com/business/news/virginia-tech-planning—acre-automation-park-to-test-drones/article_56e245ff-af92-54a4-91ac-da839cc327f2.html

Di John, Jonathan. "Is There Really a Resource Curse? A Critical Survey of Theory and Evidence." *Global Governance* 17, no. 2 (2011): 167–84.

Dunaway, Wilma. *The First American Frontier: Transition to Capitalism in Southern Appalachia, 1700–1860*, The Fred W. Morrison Series in Southern Studies. Chapel Hill: The University of North Carolina Press, 1996.

Duncan, Cynthia. *Worlds Apart: Why Poverty Persists in Rural America*. New Haven, CT: Yale University Press, 1999.

Einsiedel, Edna F., Erling Jelsøe, and Thomas Breck. "Publics at the Technology Table: The Consensus Conference in Denmark, Canada, and Australia." *Public Understanding of Science* 10, no. 1 (2001): 83–89.

Epstein, Alex. *The Moral Case for Fossil Fuels*. New York: Portfolio, 2014.

Ezrahi, Yaron. *The Descent of Icarus: Science and the Transformation of Contemporary Democracy*. Cambridge, MA: Harvard University Press, 1990.

Feenberg, Andrew. "What Is Philosophy of Technology?" In *Defining Technological Literacy: Towards an Epistemological Framework*, edited by J.R. Dakers. 5–16. New York: Palgrave Macmillan, 2006.

Feenberg, Andrew, and Michel Callon. *Between Reason and Experience: Essays in Technology and Modernity*. Cambridge, MA: The MIT Press, 2010.

Felton, Ed, and Terah Lyons. *The Administration's Report on the Future of Artificial Intelligence* (blog), October 12, 2016. https://obamawhitehouse.archives.gov/blog/2016/10/12/administrations-report-future-artificial-intelligence

Feyerabend, Paul. *Against Method: Outline of an Anarchistic Theory of Knowledge*. 4th ed. New York: Verso, 2010.

Fisher, Stephen L., and Barbara Ellen Smith, ed. *Transforming Places: Lessons from Appalachia*. Champaign-Urbana: University of Illinois Press, 2012.

Fitzgerald, Michael. *Splendid Failure: Postwar Reconstruction in the American South*. American Ways Series. Chicago: Ivan R. Dee, 2007.

Foley, Jonathan. *A five-step plan to feed the world.* Feeding 9 Billion—National Geographic. 2015. Accessed December 11, 2022, from https://www.nationalgeographic.com/foodfeatures/feeding-9-billion/

"Former Energy Center Director Prospects for Appalachian Opportunity." Virginia Tech News | Virginia Tech, February 3, 2022. https://news.vt.edu/articles/2022/01/former-energy-center-director-turns-attention-to-revitalizing-ap.html.

Foucault, Michel. *Discipline and Punish: The Birth of the Prison*. New York: Vintage Books, 1977.

Foucault, Michel. *Society, Territory, Population. Lectures at the Collège de France, 1977–78*. Edited by Michel Senellart, Francois Ewald and Alessandro Fontana. Translated by Graham Burchell. Editions du Seuil/Gallimard 2004; London: Picador, 2007.

Fox, Thomas R., Eric J. Jokela, H. Lee Allen, "The evolution of pine plantation silviculture in the Southern United States," In Gen. Tech. Rep. SRS 75, Asheville, NC: U.S. Department of Agriculture, Forest Service, Southern Research Station (2004), Chapter 8: 63–82.

Freudenburg, William R., "Addictive Economies: Extractive Industries and Vulnerable Localities in a Changing World Economy," *Rural Sociology* 57 no. 3 (Fall 1992): 305–32, https://doi.org/10.1111/j.1549-0831.1992.tb00467

Fueller Johann, Hans Muehlbacher, Kurt Matzler, and Gregor Jawecki, "Consumer Empowerment through Internet Based Co-creation." *Journal of Management Information Systems* 26, no. 3 (Winter 2009–2010): 71–102.

Gade, Eric. "Rebuilding the Ark: Alex Epstein and the Moral Case for Fossil Fuels." *BLOG// Los Angeles Review of Books*. August 4, 2017, accessed on July 13, 2023, https://blog.lareviewofbooks.org/reviews/rebuilding-ark-alex-epstein-moral-case-fossil-fuels/

Gaventa, John. *Power and Powerlessness: Quiescence and Rebellion in an Appalachian Valley*. Chicago: University of Illinois Press, 1982.

Gieryn, Thomas. "Boundary-Work and the Demarcation of Science from Non-Science: Strains and Interests in Professional Ideologies of Scientists." *American Sociological Review* 48, no. 6 (1983): 781–95.

Giroux, Henry. "Neoliberalism, corporate culture, and the promise of Higher Education: The University as a Democratic Public Sphere." *Harvard Educational Review* 72, no. 4. (December 1, 2002): 425–64. https://doi.org/10.17763/haer.72.4.0515nr62324n71p1

Graeber, Daniel J., "Statoil Completes Sale of Marcellus Acreage," *Home / Energy News*, September 16, 2016. https://www.upi.com/Statoil-completes-sale-of-Marcellus-acreage/1601474030319/

Gross, Rachel E. "The Moral Cost of Cats." *Smithsonian.com*. September 20, 2016. Retrieved December 11, 2022, from https://www.smithsonianmag.com/science-nature/moral-cost-of-cats-180960505/

Grunwald, Michael. "Trump's Love Affair with Coal." *Politico* (October 15, 2017). https://www.politico.com/magazine/story/2017/10/15/trumps-love-affair-with-coal-215710

Haraway, Donna. *Simians, Cyborgs and Women*: *The Reinvention of Nature*. New York: Routledge, 1991.

Harrington, Michael. *The Other America*: *Poverty in the United States*. New York: Macmillan, 1962.

Harvey, David. *A Brief History of Neoliberalism*. New York: Oxford University Press, 2007.

Hecht, Gabrielle. "Rupture Talk in the Nuclear Age: Conjugating Colonial Power in Africa." *Social Studies of Science* 32, no. 5/6, (2002): 691–727.

Hedges, Chris, and Joe Sacco. *Days of Destruction*, *Days of Revolt*. New York: Nation Books, 2012

Hendryx, Michael and Melissa M. Ahern, "Relations Between Health Indicators and Residential Proximity to Coal Mining in West Virginia," *American Journal of Public Health*, April 2008, 669–71.

"Dr. Michael Hendryx, Measuring Mining's Toll on Health." PRI's Environmental News Magazine. Living On Earth. Boston, MA: Public Radio International, August 11, 2012. http://www.loe.org/blog/blogs.html/?seriesID=1&blogID=17

High, Steven. "Capital and Community Reconsidered: The Politics and Meaning of Deindustrialization." *Labour/Le Travail* 55 (2005): 191–200.

Hodge, Dan. "Appalachian Coal Industry, Power Generation, and Supply Chain." By Hodge Economic Consulting, *Appalachian Regional Commission* (March 16, 2016), https://www.arc.gov/report/www-arc-gov-power/

Holm Jr., Edwin Emanuel. *Production and Marketing of Coal in Virginia and the Nation*. Richmond: Virginia Division of Industrial Development and Planning, 1955.

Holpuch, Amanda, and Nadja Popvich. "Rural Counties across the US Becoming a Powder Keg for HIV Outbreak." *The Guardian*. July 31, 2016. https://www.theguardian.com/society/2016/jul/31/rural-us-counties-hiv-outbreak-cdc

Hook, Janet and Kris Maher, "Don Blankenship, Former Coal CEO, to Run for West Virginia Senate," Wall Street Journal, last updated November 29, 2017, 8:08PM

ET https://www.wsj.com/articles/former-coal-executive-convicted-on-charges-tied-to-mine-explosion-to-run-for-senate-1511987189

Horowitz, Roger, *Boys and Their Toys: Masculinity, Class and Technology in America*. Hagley Perspectives on Business and Culture. New York: Routledge, 2001.

Horton, Myles, Paulo Freire, Brenda Bell, John Gaventa, and John Peters. *We Make the Road by Walking: Conversations on Education and Social Change*. Temple University Press, 1990.

Hulme, Mike. *Why We Disagree about Climate Change*. New York: Cambridge University Press, 2009.

Indian Ministry of Rural Development, "National Rural Employment Guarantee Act, 2005." n.d. https:rural.nic.in/documents/policies-acts-bills/department-rural-development. Accessed on July 13, 2023.

Jaramillo, Paulina, Michael Griffin, and Scott Matthews. "Comparative Life-Cycle Air Emissions of Coal, Domestic Natural Gas, LNG, and SNG for Electricity Generation." *Environmental Science & Technology* 41, no. 17 (2007): 6290–96.

Jasanoff, Shelia. *States of Knowledge: The Co-Production of Science and the Social Order*. International Library of Sociology. New York: Routledge, 2004.

Jasanoff, Sheila. *The Fifth Branch: Science Advisers as Policymakers*. Cambridge, MA: Harvard University Press, 2009.

Joel, Billy, "Allentown," *The Nylon Curtain* (September 1982).

Joerges, Bernward. "Large Technical Systems and the Discourse of Complexity." In *Complex Technical Systems*, edited by Lars Ingelstam. Stockholm: Affärs Litteratur, 1996: 55–72.

Johnson, Kenneth. "Demographic Trends in Rural and Small Town America." The Carsey School of Public Policy at the Scholars' Repository, March 15, 2006. http://scholars.unh.edu/carsey/5

Johnston, A. Edward, Paul R. Poulton, and Kevin Coleman. "Chapter 1 Soil Organic Matter: Its Importance in Sustainable Agriculture and Carbon Dioxide Fluxes." *Advances in Agronomy* 101 (2009): 1–57.

King, Stephen M., Bradley S. Chilton, and Gary E. Roberts. "Reflections on Defining the Public Interest." *Administration & Society* 41, no. 8 (November 6, 2009): 954–78. https://doi.org/10.1177/0095399709349910

Kohler, Robert E. *Landscapes and Labscapes: Exploring the Lab-Field Border in Biology*. Chicago: The University of Chicago Press, 2002.

Kopittke, Peter M., Neal W. Menzies, Peng Wang, Brigid A. McKenna, and Enzo Lombi. "Soil and the intensification of Agriculture for Global Food Security." *Environment International* 132 (November 2019): 105078. https://doi.org/10.1016/j.envint.2019.105078

Krishna, Santosh, Suzanne Austin Boren, and E. Andrew Bates. "Healthcare via Cell Phones: A Systematic Review." *Telemedicine and E-Health* 15, no. 3 (April 2009): 231–40.

Kuhn, Thomas. *Structure of Scientific Revolutions*. Chicago: University of Chicago Press, 1962.

Kurz, Tim, Shona Crabb, and Martha Augoustinos. "Contesting the 'National Interest' and Maintaining 'Our Lifestyle': A Discursive Analysis of Political Rhetoric around Climate Change." *British Journal of Social Psychology* 49, no. 3 (n.d.): 601–25.

Larsen, Erling Roed. "Escaping the Resource Curse and the Dutch Disease? : When and Why Norway Caught Up with and Forged ahead of Its Neighbors." Statistics Norway, Research Department, 2004. http://hdl.handle.net/11250/180569

LaRose, Robert, Jennifer Gregg, Sharon Stover, Joseph Straubhaar, and Serena Carpenter. "Closing the Rural Broadband Gap: Promoting Adoption of the Internet in Rural America." *Telecommunications Policy* 31, no. 6 (2007): 359–73.

Latour, Bruno. "On Actor-Network Theory: A Few Clarifications." *Soziale Welt* 47 (1996): 369–81.

Latour, Bruno. *Science in Action: How to Follow Scientists and Engineers through Society*. Cambridge, MA: Harvard University Press, 1987.

Latour, Bruno, and Steve Woolgar. *Laboratory Life: The Construction of Scientific Facts, Second Edition*. Princeton: Princeton University Press, 1986.

Leadbetter, David, ed. *Mining Town Crisis: Globalization, Labour and Resistance in Sudbury*. Black Point: Fernwood Publishing Co., Ltd., 2008.

Leadbetter, David. "Single-Industry Resource Communities: 'Shrinking,' and the New Crisis of Hinterland Economic Development." In *The Future of Shrinking Cities: Problems, Patterns and Strategies of Urban Transformation in a Global Context*, edited by Karina Pallagst, et al. 89–100. Berkeley: Center for Global Metropolitan Studies, Institute of Urban and Regional Development and the Shrinking Cities International Research Network, 2009.

Ledet, Richard. "Correlates of Corruption." *Public Integrity* (Routledge) 13, no. 2 (2011): 149–62.

Lee, Kristin. "Artificial Intelligence, Automation, and the Economy." *What's Happening* (blog), December 20, 2016, National Archives and Records Administration. Retrieved July 29, 2023, from https://obamawhitehouse.archives.gov/blog/2016/12/20/artificial-intelligence-automation-and-economy

Lepschy, Giulio. *A Survey of Structural Linguistics*. London: Faber and Faber Limited, 1970.

Letson, Al. "Appalachia Rising." States of the Re:Union, 2011. (Site discontinued.)

Lewis, Ronald L. *Black Coal Miners in America: Race, Class, and Community Conflict, 1780–1980*. Lexington: University Press of Kentucky, 1987.

Lispector, Clarice, and Giovanni Pontiero. *Selected cronicas*. New Directions Pub. Corp., 1996.

Locklear, Erica Abrams. "A Matter of Taste: Reading Food and Class in Appalachian Literature." Paper presented at Thirty-Fourth Annual Appalachian Studies Conference, University of North Carolina at Asheville, 2011. However, it has since been published.

Locklear, Erica Abrams. "A Matter of Taste: Reading Food and Class in Appalachian Literature." In *Writing in the Kitchen: Essays on Southern Literature and Foodways*, edited by David A. Davis and Tara Powell, 124–42. University Press of Mississippi, 2014.

Lovins, Amory. B. "Non-Nuclear Energy Strategies." *International Relations* 5, no. 3 (April 1976): 998–1010. https://doi.org/10.1177/004711787600500302

Lowe, Travis. "I've Seen The Future. It Looks Like Appalachia: Life after the Job Apocalypse." *Medium* (blog), March 31, 2017. https://medium.com/startup-grind/ive-been-to-the-future-it-looks-like-appalachia-5553bf89f275

Martinez-Fernandez, Cristina and Chung-Tong Wu. "Shrinking Cities: A Global Overview and Concerns about Australian Mining Cities." In *The Future of Shrinking Cities: Problems, Patterns and Strategies of Urban Transformation in a Global Context*, edited by Karina Pallagst, et al. Berkeley: Center for Global Metropolitan Studies, Institute of Urban and Regional Development and the Shrinking Cities International Research Network, 2009, 29–36.

Martin, Richard. *Coal Wars: The Future of Energy and the Fate of the Planet*. New York: St. Martin's Press, 2015.

Mbembe, Achille. "Necropolitics." trans. Libby Meintjes, *Public Culture* 15, no. 1 (Winter 2003): 11–40.

McGuire, Mary Richie. "The Chesapeake: Bacon's Promise, Boyle's Project' in 'Translating Natural Knowledge in the Age of Revolution: Tobacco, Science, and the Rights of Man and Nature in the Journals of Benjamin Henry Latrobe, 1795–1820." PhD dis.., Virginia Tech, 2018.

Menand, Louis. *The Marketplace of Ideas: Reform and Resistance in the American University*. New York: W.W. Norton, 2010.

Merchant, Carolyn. *The Death of Nature: Women Ecology, and the Scientific Revolution*. New York: Harper and Row, 1983.

Merton, Robert K. *The Sociology of Science: Theoretical and Empirical Investigations*. Chicago: The University of Chicago Press, 1972.

Mitchell, Timothy. *Carbon Democracy: Political Power in the Age of Oil* (New York: Verso, 2011).

Moxley, T. (2022, February 3). Former Energy Center Director Prospects for Appalachian Opportunity. VTx. Retrieved December 23, 2022, from https://vtx.vt.edu/articles/2022/01/former-energy-center-director-turns-attention-to-revitalizing-ap.html

Mulkay, Michael J. "Norms and Ideology in Science." *Social Science Information* 15, no. 4–5, 1976.

Muro, Mark and Bruce Katz. *The New 'Cluster' Moment: How Regional Innovation Clusters Can Foster the Next Economy*. Metropolitan Policy Program. The Brookings Institution, September 2010.

Nelkin, Dorothy. *Controversy: Politics of Technical Decisions*. Beverly Hills: SAGE Publications, 1992.

Nelkin, Dorothy. *Dangerous Diagnostics: The Social Power of Biological Information*. Chicago: The University of Chicago Press, 1994.

Nelkin, Dorothy. "The Political Impact of Technical Expertise." *Social Studies of Science* 5, no. 1 (1975): 35–54.

Office of Fossil Energy and Carbon Management (FECM), "University Training and Research (UTR)," n.d., updated to include 2018 information. Accessed

July 25, 2023. https://www.energy.gov/fecm/university-training-and-research-utr-program

Okin, Gregory S. "Environmental Impacts of Food Consumption by Dogs and Cats." *PLOS ONE* 12, no. 8. (August 2, 2017). https://doi.org/10.1371/journal.pone.0181301

"Organization: About Us." U.S. Agency for International Development, January 17, 2023.

https://www.usaid.gov/about-us/organization.

"Outbreak of E. Coli Infections Linked to Romaine Lettuce." Centers for Disease Control and Prevention. Centers for Disease Control and Prevention, January 9, 2019. https://www.cdc.gov/ecoli/2018/o157h7-11-18/index.html

Pallagst, Karina. "Shrinking Cities in the United States of America: Three Cases, Three Planning Stories." In *The Future of Shrinking Cities: Problems, Patterns and Strategies of Urban Transformation in a Global Context*, edited by Karina Pallagst et al. Berkeley: Center for Global Metropolitan Studies, Institute of Urban and Regional Development and the Shrinking Cities International Research Network, 2009: 81–88.

Pancake, Ann. *Strange as this Weather Has Been: a Novel.* Emeryville, CA, 2007.

Parker, Laura. *Parched: A New Dust Bowl Forms in the Heartland.* National Geographic Food. May 3, 2021. Retrieved December 11, 2022, https://www.nationalgeographic.com/science/article/140516-dust-bowl-drought-oklahoma-panhandle-food

Parker Suzi. "How Poultry Producers Are Ravaging the RURAL SOUTH." *Grist Magazine* (February 22, 2006), http://grist.org/article/parker1/

Parker, William N. "Introduction: The Cotton Economy of the Antebellum South." *Agricultural History* 44, no. 1 (1970): 1–4.

Perry, Huey. *"They'll Cut Off Your Project": a Mingo County Chronicle.* New York: Praeger Publishers, 1972.

Pohjanpalo, Kati. "Finland Investigating Russia for GPS Jamming: 'Civilian Flights Have Been Put in Danger." *Bloomberg News*, November 18, 2018, https://nationalpost.com/news/world/finland-investigating-russia-for-gps-jamming-civilian-flights-have-been-put-in-danger

Popper, Karl R. *The Open Society and Its Enemies, Vol. 1, The Spell of Plato.* New Jersey: Princeton University Press, 1966.

Popper, Karl R. *The Open Society and Its Enemies, Vol. 2, The High Tide of Prophecy: Hegel, Marx, and the Aftermath.* New Jersey: Princeton University Press, 1966.

Porteri, Corinna. "Sergio Sismondo: An Introduction to Science and Technology Studies." *Theoretical Medicine and Bioethics* 33 (October 1, 2011): 233–35, https://doi.org/10.1007/s11017-011-9197-6

Pretenders, The, "My City Was Gone," B-side to *Back on the Chain Gang (Single)*, (October 1982).

Rafferty, Matthew. "The Bayh–Dole Act and University Research and Development." *Research Policy* 37, no. 1 (2008): 29–40. https://doi.org/10.1016/j.respol.2007.06.010

Ravetz, Jerome R. "The Post-Normal Science of Precaution." *Water Science and Technology* 52, no. 5, (September 2005). https://doi.org/10.2166/wst.2005.0145

Ravetz, Jerome R. *Scientific Knowledge and Its Social Problems*. Oxford: Clarendon Press, 1971.

Reguero, Borja G., Iñigo J. Losada, Fernando J. Méndez. "A Recent Increase in Global Wave Power as a Consequence of Oceanic Warming." *Nature Communications* 10, no. 1 (January 14, 2019). https://doi.org/10.1038/s41467-018-08066-0

Reichert Powell, Douglas, *Critical Regionalism: Connecting Politics and Culture in an American Landscape*. Chapel Hill: University of North Carolina Press, 2007.

Reid, Herbert, and Betsy Taylor. "Appalachia as a Global Region: Toward a Critical Regionalism and Civic Professionalism." *Journal of Appalachian Studies* 8, no. 1 (Spring 2002): 9–32.

Reid, Herbert, and Betsy Taylor. *Recovering the Commons: Democracy, Place, and Global Justice*. Board of Trustees of the University of Illinois, 2010.

Renner, Michael, Celia Garcia-Banos, and Arslan Khalid, *Renewable Energy and Jobs: Annual Review 2022*, IRENA. https://policycommons.net/artifacts/2685048/renewable-energy-and-jobs/

Reuters Staff. "China Energy Investment Signs MOU for $83.7 Billion in West Virginia Projects," *Reuters*, November 9, 2017. http://www.reuters.com/article/us-trump-asia-energy-west-virginia/china-energy-investment-signs-mou-for-83-7-billion-in-west-virginia-projects-idUSKBN1D90S9?il=0

Rich, Nathaniel. "The Most Honest Book about Climate Change Yet." Review of *Carbon Ideologies* by William T. Vollmann, *The Atlantic*. Atlantic Media Company (September 9, 2018). https://www.theatlantic.com/magazine/archive/2018/10/william-vollmann-carbon-ideologies/568309/

Rodrik, Dan and Arvind Subramanian and Francesco Trebbi. "Institutions Rule: The Primacy of Institutions over Geography and Integration in Economic Development," *Journal of Economic Growth* 9, no. 2 (June 2004): 131–65.

Rogers, Everett M., "The Intellectual Foundation and History of the Agricultural Extension Model," *Science Communication* 9, no. 4 (1988): 492–510. https://doi.org/10.1177/0164025988009004003

Roochnik, David. "Socrates's Use of the Techne-Analogy." *Journal of the History of Philosophy* 24, no. 3 (1986): 295–310.

Rothschild, Joyce. "Workers' Cooperatives and Social Enterprise: A Forgotten Route to Social Equity and Democracy." *American Behavioral Scientist* 52, no. 7 (2009): 1023–41.

Rustin, Michael. "The Neoliberal University and its Alternatives." *Soundings* no. 63 (July 2016): 147–76. https://doi.org/10.3898/136266216819377057

Sakolski, Aaron. M. *The Great American Land Bubble: The Amazing Story of Land-Grabbing, Speculations, and Booms from Colonial Days to Present Time*. New York: Harpers & Brothers, 1966.

Satterwhite, Emily. *Dear Appalachia: Readers, Identity and Popular Fiction since 1878*. Lexington: University Press of Kentucky, 2011.

Schafft, Kai A., and Alicia Youngblood Jackson. *Rural Education for the Twenty-First Century: Identity, Place, and Community in a Globalizing World*. University Park, PA: Pennsylvania State University Press, 2010.

Schumpeter, Joseph. *Capitalism, Socialism and Democracy*, 1947, Reprint, New York: Routledge, 2010.

Scott, James C. *Seeing Like A State: How Certain Schemes to Improve the Human Condition Have Failed*. New Haven, CT: Yale University Press, 1998.

Scott, Rebecca. *Removing Mountains: Extracting Nature and Identity in the Appalachian Coalfields*. Quadrant. Minneapolis: University of Minnesota Press, 2010.

Shapin, Steven, and Simon Schaeffer. *Leviathan and the Air-Pump: Hobbes, Boyle, and the Experimental Life*. Princeton University Press, 1985.

Shuman, Michael H. *The Small Mart Revolution: How Local Businesses Are Beating the Global Competition*. San Francisco: Bernett-Koehler, 2006.

Smith, Barbara Ellen. "De-Gradations of Whiteness: Appalachia and the Complexities of Race." *Journal of Appalachian Studies* 10, no. 1/2 (Spring–Fall 2004): 38–57.

Smith, Barbara E., ed. *Neither Separate Nor Equal*. Temple University Press, 1999.

Smith, Merrit Roe & Leo Marx, *Does Technology Drive History? The Dilemma of Technological Determinism*. Cambridge, MA: The MIT Press, 1994.

Spencer, L. "Decades Will Be Needed to Fully Assess Fracking's Impact on Drinking Water, Geological Survey Says," The Rural Blog, University of Kentucky, Institute for Rural Journalism and Community Issues, June 1, 2017, accessed July 21, 2023, https://irjci.blogspot.com/search?q=decades+will+be+needed+to+fully+assess

Srinivasan, Prianka. "Appalachia's Fickle Friend." *Jacobin*. December 5, 2017. https://jacobin.com/2017/12/appalachia-friends-of-coal-mining-industry-jobs-environmental-just-transition

Stafford, Thomas. *Afflicting the Comfortable: Journalism and Politics in West Virginia*. Morgantown, WV: West Virginia University Press, 2005.

Stanley, Talmage. *The Poco Field: An American Story of Place* (Urbana and Chicago: University of Illinois Press. 2012).

Stanley, Talmage. "The Poco Field: Politics, Culture, and Place in Contemporary Appalachia."

PhD diss., Emory University; 1996, ProQuest Dissertations Publishing.

Stephenson, Garry, Larry Lev, and Linda Brewer. "'I'm Getting Desperate': What We Know about Farmers' Markets That Fail." *Renewable Agriculture and Food Systems* 23, no. 3 (2008): 188–99. https://doi.org/10.1017/S1742170507002153

Stirling, Andrew. *On the Economics and Analysis of Diversity*. Brighton, UK: University of Sussex, Science Policy Research Unit, 1998.

Stoll, Jennifer, Jessie P. H. Poon, and Trina Hamilton. "Sustainable Practice? An Examination of Canada's Agricultural and Energy Cooperatives." *Professional Geographer* 67, no. 2 (2015): 187–94.

Stone, Deborah. *Policy Paradox: The Art of Political Decision Making*. New York: Norton, 2001.

Sweeten, Sara E., J. Sweeten, John R. Craynon, William M. Ford, and Stephen H. Schoenholtz. "Evaluation of Current Measures of Aquatic Biological Integrity in the Central Appalachian Coalfields: Efficacy and Implications." In Environmental

Considerations in Energy Production, edited by John R. Craynon, 381–94. Englewood: Society for Mining Metallurgy. and Exploration Inc., 2013.

Tewalt, Susan, Jason Willett, and Robert Finkelman. "The World Coal Quality Inventory: A Status Report." *International Journal of Coal Geology*.63, no. 1/2 (July 2005): 190–94.

"The Global Environmental Paw Print of Pet Food." *Global Environmental Change* 65, (November 2020): 102153. https://doi.org/10.1016/j.gloenvcha.2020.102153

Thomas, Jerry Bruce. *An Appalachian Reawakening: West Virginia and the Perils of the New Machine Age, 1945–1972. Vol. 12*, Morgantown: West Virginia University Press, 2010.

Thomas, Jerry Bruce, "Jedediah Hotchkiss, Gilded-Age Propagandist of Industrialism," *The Virginia Magazine of History and Biography* 84, no. 2 (April 1976): 189–202.

Thompson, Richard C., Charles J. Moore, Frederick S vom Saal, and Shanna H. Swan. "Plastics, the Environment and Human Health: Current Consensus and Future Trends." *Philosophical Transactions of the Royal Society of London, Series B, Biological Sciences* 364, no. 1526 (July 27, 2009): 2153–66. https://doi.org/10.1098/rstb.2009.0053. PMID: 19528062. PMCID: PMC2873021

Thyberg, Krista L., and David J. Tonjes. "Drivers of Food Waste and Their Implications for Sustainable Policy Development." *Resources, Conservation and Recycling* 106 (January 2016): 110–23. https://doi.org/10.1016/j.resconrec.2015.11.016

Tidd, Joe, John Bessant, and Keith Pavitt. *Managing Innovation: Integrating Technology, Market and Organizational Change, 3rd Edition*. Chichester, UK: John Wiley & Sons Ltd, 2005.

Tierney, John. "A Gift from the '70s: Energy Lessons." *The New York Times*, October 7, 2008. https://www.nytimes.com/2008/10/07/science/07tier.html

Tolbert, Charles M., Michael D. Irwin, Thomas A. Lyson, and Alfred R. Nucci. "Civic Community in Small-Town America: How Civic Welfare Is Influenced by Local Capitalism and Civic Engagement." *Rural Sociology* 67, no. 1 (March 1, 2002): 90–113.

Trotter, Joe. *Coal, Class, and Color: Blacks in Southern West Virginia, 1915–32*. Blacks in the New World. Chicago: University of Illinois Press, 1990.

Tucker, Spencer C. *Brigadier General John D. Imboden: Confederate Commander in the Shenandoah*. Lexington: University Press of Kentucky, 2003.

Turrentine, Jeff. "Coal Is Literally Killing Us," NRDC, June 30, 2016, accessed on July 13, 2023, https://www.nrdc.org/onearth/coal-literally-killing-us

Union of Concerned Scientists. *The Natural Gas Gamble: A Risky Bet on America's Clean Energy Future* (March 4, 2015). http://www.ucsusa.org/clean-energy/coal-and-other-fossil-fuels/natural-gas-gamble-risky-bet-on-clean-energy-future#.WeMjKxmGPrc

U.S. Census Bureau, "Metropolitan—Micropolitan," Programs—Surveys, July 13, 2023, accessed July 13, 2023, https://www.census.gov/programs-surveys/metro-micro.html

U.S. Department of Agriculture, Economic Research Service, "Farm Labor," updated March 22, 2023, https://www.ers.usda.gov/topics/farm-economy/farm-labor/

U.S. Department of Labor, Mine Safety and Health Administration, "History of Mine Safety and Health Legislation," n.d. Accessed July 19, 2023, https://arlweb.msha.gov/MSHAINFO/MSHAINF2.htm.

U.S. Energy Information Administration, "U.S. Energy Facts Explained," June 10, 2022 accessed July 13, 2023, https://www.eia.gov/energyexplained/?page=us_energy_home)

U.S. Environmental Protection Agency, Science Advisory Board, *EPA-SAB-11-006*: *Review of Field-Based Aquatic Life Benchmark for Conductivity in Central Appalachian Streams*, https://archive.epa.gov/water/archive/web/pdf/epa-sab-11-006-unsigned.pdf

U.S. Geological Survey, "Methane and Benzene in Drinking-Water Wells Overlying the Eagle Ford, Fayetteville, and Haynesville Shale Hydrocarbon Production Areas" by Peter B. McMahon, Jeannie R.B. Barlow, Mark A. Engle, Kenneth Belitz, Patricia B. Ging, Andrew G. Hunt, Bryant C. Jurgens, Yousif K. Kharaka, Roland W. Tollett, and Timothy M. Kresse, *Environmental Science and Technology* 51, no. 12 (May 31, 2017): 6727–34, https://doi.org/10.1021/acs.est.7b00746

Valencia, Richard. *The Evolution of Deficit Thinking*: *Educational Thought and Practice*. *Stanford Series on Education & Public Policy*. New York: Routledge, 1997.

Von Hippel, Eric. *Democratizing Innovation*. Cambridge, MA: MIT Press, 2005.

Von Hippel, Eric. "Horizontal Innovation Networks-by and for Users." *Industrial & Corporate Change* 16, no. 2 (April 2007): 293–315.

Wainaina, Binyavanga. "How to Write about Africa." (2005): 4.

Wallace, Anthony F. C. *Saint Clair*: *A Nineteenth Century Coal Town's Experience with a Disaster-Prone Industry*. New York: Random House, 1987.

Walker, Frank X. "Affrilachia. Lexington." (2000).

Ward, Jr., Ken, Coal Tattoo (blog), http://blogs.wvgazettemail.com/coaltattoo/

Warren, Mark E. "What Does Corruption Mean in a Democracy?" *American Journal of Political Science*, 48, no. 2 (2010): 328–43.

Weiner, Deborah. *Coalfield Jews*: *An Appalachian History*. Chicago: University of Illinois Press, 2006.

West Virginia Department of Health and Human Resources, "West Virginia Experiences Increase in Overdose Deaths; Health Officials Emphasize Resources," *2021 News and Announcements*, April 23, 2021, accessed, July 20, 2023: https://dhhr.wv.gov/News/2021/Pages/West-Virginia-Experiences-Increase-in-Overdose-Deaths;-Health-Officials-Emphasize-Resources.aspx

Whitt, Christine, Jessica E. Todd, and James M. MacDonald. *America's Diverse Family Farms*: *2020 Edition*. Economic Information Bulletin No. (EIB-220), U.S. Department of Agriculture, Economic Research Service, December 2020. https://www.ers.usda.gov/publications/pub-details/?pubid=100011.

Williams, Jr. Hank, "A Country Boy Can Survive" (January 1982).

Williams, Travis L. "Corporate Accountability and Environmental Health Advocacy in Silicon Valley." Paper presented at the Dimensions of Political Ecology Conference, Lexington, KY, February 2014.

Winner, Langdon. *The Whale and the Reactor*: *A Search for Limits in an Age of High Technology*. Chicago: University of Chicago Press, 1986.

Woolley, Shannon M., Susan L. Meacham, Lauren C. Balmert, Evelyn O. Talbott, and Jeanine M. Buchanich, "Comparison of Mortality Disparities in Central Appalachian Coal- and Non-Coal-Mining Counties," *Journal of Occupational and Environmental Health* 57, no. 6 (June 2015): 687–94.

World Nuclear Association. "'Clean Coal' Technologies, Carbon Capture & Sequestration." (updated November 2021). http://www.world-nuclear.org/information-library/energy-and-the-environment/clean-coal-technologies.aspx

Wvustatler. "Mining Engineering: Research." Mining Engineering | West Virginia University. Accessed August 20, 2023. http://mine.statler.wvu.edu/research.

Zahran, Sammy, Samuel D. Brody, Arnold Vedlitz, Michael G. Lacy, and Chelsea Lynn Schelly. "Greening Local Energy: Explaining the Geographic Distribution of Household Solar Energy Use in the United States." *Journal of the American Planning Association* 74, no. 4 (2008), 419–34.

Zinn, Howard. *A People's History of the United States*. New York: Harper Perennial Modern Classics, 2005.

Index

About the Author

Crystal Cook Marshall grew up around cows and coal in West Virginia and Virginia. A study-abroad award to Germany set her on a path to work in the United States and internationally in education and nonprofit/NGOs, where she led institutional creation and development, strategy, systems analysis, and policy in gender equity in Hollywood, democratic civil sector support in the South Caucasus, and economic sector revitalization in Appalachia. Cook Marshall is a Barnard College alumna with master's degrees from the New School and Antioch, and a PhD in science and technology studies from Virginia Tech. A Fulbright alumna in writing and a National Science Foundation Graduate Fellow in sociology and policy of science and technology, Cook Marshall currently works on issues of climate change in the rural. Cook Marshall also farms in the foothills of North Carolina and is in process on a community space in McDowell County, West Virginia. She is a member of the faculty at North Carolina A&T State University.